FREE Test Taking Tips DVD Offer

To help us better serve you, we have developed a Test Taking Tips DVD that we would like to give you for FREE. **This DVD covers world-class test taking tips that you can use to be even more successful when you are taking your test.**

All that we ask is that you email us your feedback about your study guide. Please let us know what you thought about it – whether that is good, bad or indifferent.

To get your **FREE Test Taking Tips DVD**, email freedvd@studyguideteam.com with "FREE DVD" in the subject line and the following information in the body of the email:

a. The title of your study guide.

b. Your product rating on a scale of 1-5, with 5 being the highest rating.

c. Your feedback about the study guide. What did you think of it?

d. Your full name and shipping address to send your free DVD.

If you have any questions or concerns, please don't hesitate to contact us at freedvd@studyguideteam.com.

Thanks again!

2—

ISEE Lower Level Test Prep

Three Lower Level ISEE Practice Tests for the Independent School Entrance Exam

[2nd Edition Book]

TPB Publishing

Written and edited by TPB Publishing.

ISBN 13: 9781628456622
ISBN 10: 1628456620

Table of Contents

Quick Overview

As you draw closer to taking your exam, effective preparation becomes more and more important. Thankfully, you have this study guide to help you get ready. Use this guide to help keep your studying on track and refer to it often.

This study guide contains several key sections that will help you be successful on your exam. The guide contains tips for what you should do the night before and the day of the test. Also included are test-taking tips. Knowing the right information is not always enough. Many well-prepared test takers struggle with exams. These tips will help equip you to accurately read, assess, and answer test questions.

A large part of the guide is devoted to showing you what content to expect on the exam and to helping you better understand that content. In this guide are practice test questions so that you can see how well you have grasped the content. Then, answer explanations are provided so that you can understand why you missed certain questions.

Don't try to cram the night before you take your exam. This is not a wise strategy for a few reasons. First, your retention of the information will be low. Your time would be better used by reviewing information you already know rather than trying to learn a lot of new information. Second, you will likely become stressed as you try to gain a large amount of knowledge in a short amount of time. Third, you will be depriving yourself of sleep. So be sure to go to bed at a reasonable time the night before. Being well-rested helps you focus and remain calm.

Be sure to eat a substantial breakfast the morning of the exam. If you are taking the exam in the afternoon, be sure to have a good lunch as well. Being hungry is distracting and can make it difficult to focus. You have hopefully spent lots of time preparing for the exam. Don't let an empty stomach get in the way of success!

When travelling to the testing center, leave earlier than needed. That way, you have a buffer in case you experience any delays. This will help you remain calm and will keep you from missing your appointment time at the testing center.

Be sure to pace yourself during the exam. Don't try to rush through the exam. There is no need to risk performing poorly on the exam just so you can leave the testing center early. Allow yourself to use all of the allotted time if needed.

Remain positive while taking the exam even if you feel like you are performing poorly. Thinking about the content you should have mastered will not help you perform better on the exam.

Once the exam is complete, take some time to relax. Even if you feel that you need to take the exam again, you will be well served by some down time before you begin studying again. It's often easier to convince yourself to study if you know that it will come with a reward!

Test-Taking Strategies

1. Predicting the Answer

When you feel confident in your preparation for a multiple-choice test, try predicting the answer before reading the answer choices. This is especially useful on questions that test objective factual knowledge. By predicting the answer before reading the available choices, you eliminate the possibility that you will be distracted or led astray by an incorrect answer choice. You will feel more confident in your selection if you read the question, predict the answer, and then find your prediction among the answer choices. After using this strategy, be sure to still read all of the answer choices carefully and completely. If you feel unprepared, you should not attempt to predict the answers. This would be a waste of time and an opportunity for your mind to wander in the wrong direction.

2. Reading the Whole Question

Too often, test takers scan a multiple-choice question, recognize a few familiar words, and immediately jump to the answer choices. Test authors are aware of this common impatience, and they will sometimes prey upon it. For instance, a test author might subtly turn the question into a negative, or he or she might redirect the focus of the question right at the end. The only way to avoid falling into these traps is to read the entirety of the question carefully before reading the answer choices.

3. Looking for Wrong Answers

Long and complicated multiple-choice questions can be intimidating. One way to simplify a difficult multiple-choice question is to eliminate all of the answer choices that are clearly wrong. In most sets of answers, there will be at least one selection that can be dismissed right away. If the test is administered on paper, the test taker could draw a line through it to indicate that it may be ignored; otherwise, the test taker will have to perform this operation mentally or on scratch paper. In either case, once the obviously incorrect answers have been eliminated, the remaining choices may be considered. Sometimes identifying the clearly wrong answers will give the test taker some information about the correct answer. For instance, if one of the remaining answer choices is a direct opposite of one of the eliminated answer choices, it may well be the correct answer. The opposite of obviously wrong is obviously right! Of course, this is not always the case. Some answers are obviously incorrect simply because they are irrelevant to the question being asked. Still, identifying and eliminating some incorrect answer choices is a good way to simplify a multiple-choice question.

4. Don't Overanalyze

Anxious test takers often overanalyze questions. When you are nervous, your brain will often run wild, causing you to make associations and discover clues that don't actually exist. If you feel that this may be a problem for you, do whatever you can to slow down during the test. Try taking a deep breath or counting to ten. As you read and consider the question, restrict yourself to the particular words used by the author. Avoid thought tangents about what the author *really* meant, or what he or she was *trying* to say. The only things that matter on a multiple-choice test are the words that are actually in the question. You must avoid reading too much into a multiple-choice question, or supposing that the writer meant something other than what he or she wrote.

5. No Need for Panic

It is wise to learn as many strategies as possible before taking a multiple-choice test, but it is likely that you will come across a few questions for which you simply don't know the answer. In this situation, avoid panicking. Because most multiple-choice tests include dozens of questions, the relative value of a single wrong answer is small. As much as possible, you should compartmentalize each question on a multiple-choice test. In other words, you should not allow your feelings about one question to affect your success on the others. When you find a question that you either don't understand or don't know how to answer, just take a deep breath and do your best. Read the entire question slowly and carefully. Try rephrasing the question a couple of different ways. Then, read all of the answer choices carefully. After eliminating obviously wrong answers, make a selection and move on to the next question.

6. Confusing Answer Choices

When working on a difficult multiple-choice question, there may be a tendency to focus on the answer choices that are the easiest to understand. Many people, whether consciously or not, gravitate to the answer choices that require the least concentration, knowledge, and memory. This is a mistake. When you come across an answer choice that is confusing, you should give it extra attention. A question might be confusing because you do not know the subject matter to which it refers. If this is the case, don't eliminate the answer before you have affirmatively settled on another. When you come across an answer choice of this type, set it aside as you look at the remaining choices. If you can confidently assert that one of the other choices is correct, you can leave the confusing answer aside. Otherwise, you will need to take a moment to try to better understand the confusing answer choice. Rephrasing is one way to tease out the sense of a confusing answer choice.

7. Your First Instinct

Many people struggle with multiple-choice tests because they overthink the questions. If you have studied sufficiently for the test, you should be prepared to trust your first instinct once you have carefully and completely read the question and all of the answer choices. There is a great deal of research suggesting that the mind can come to the correct conclusion very quickly once it has obtained all of the relevant information. At times, it may seem to you as if your intuition is working faster even than your reasoning mind. This may in fact be true. The knowledge you obtain while studying may be retrieved from your subconscious before you have a chance to work out the associations that support it. Verify your instinct by working out the reasons that it should be trusted.

8. Key Words

Many test takers struggle with multiple-choice questions because they have poor reading comprehension skills. Quickly reading and understanding a multiple-choice question requires a mixture of skill and experience. To help with this, try jotting down a few key words and phrases on a piece of scrap paper. Doing this concentrates the process of reading and forces the mind to weigh the relative importance of the question's parts. In selecting words and phrases to write down, the test taker thinks about the question more deeply and carefully. This is especially true for multiple-choice questions that are preceded by a long prompt.

9. Subtle Negatives

One of the oldest tricks in the multiple-choice test writer's book is to subtly reverse the meaning of a question with a word like *not* or *except*. If you are not paying attention to each word in the question, you can easily be led astray by this trick. For instance, a common question format is, "Which of the following is…?" Obviously, if the question instead is, "Which of the following is not…?," then the answer will be quite different. Even worse, the test makers are aware of the potential for this mistake and will include one answer choice that would be correct if the question were not negated or reversed. A test taker who misses the reversal will find what he or she believes to be a correct answer and will be so confident that he or she will fail to reread the question and discover the original error. The only way to avoid this is to practice a wide variety of multiple-choice questions and to pay close attention to each and every word.

10. Reading Every Answer Choice

It may seem obvious, but you should always read every one of the answer choices! Too many test takers fall into the habit of scanning the question and assuming that they understand the question because they recognize a few key words. From there, they pick the first answer choice that answers the question they believe they have read. Test takers who read all of the answer choices might discover that one of the latter answer choices is actually *more* correct. Moreover, reading all of the answer choices can remind you of facts related to the question that can help you arrive at the correct answer. Sometimes, a misstatement or incorrect detail in one of the latter answer choices will trigger your memory of the subject and will enable you to find the right answer. Failing to read all of the answer choices is like not reading all of the items on a restaurant menu: you might miss out on the perfect choice.

11. Spot the Hedges

One of the keys to success on multiple-choice tests is paying close attention to every word. This is never truer than with words like almost, most, some, and sometimes. These words are called "hedges" because they indicate that a statement is not totally true or not true in every place and time. An absolute statement will contain no hedges, but in many subjects, the answers are not always straightforward or absolute. There are always exceptions to the rules in these subjects. For this reason, you should favor those multiple-choice questions that contain hedging language. The presence of qualifying words indicates that the author is taking special care with his or her words, which is certainly important when composing the right answer. After all, there are many ways to be wrong, but there is only one way to be right! For this reason, it is wise to avoid answers that are absolute when taking a multiple-choice test. An absolute answer is one that says things are either all one way or all another. They often include words like *every, always, best,* and *never*. If you are taking a multiple-choice test in a subject that doesn't lend itself to absolute answers, be on your guard if you see any of these words.

12. Long Answers

In many subject areas, the answers are not simple. As already mentioned, the right answer often requires hedges. Another common feature of the answers to a complex or subjective question are qualifying clauses, which are groups of words that subtly modify the meaning of the sentence. If the question or answer choice describes a rule to which there are exceptions or the subject matter is complicated, ambiguous, or confusing, the correct answer will require many words in order to be expressed clearly and accurately. In essence, you should not be deterred by answer choices that seem excessively long. Oftentimes, the author of the text will not be able to write the correct answer without offering some qualifications and modifications. Your job is to read the answer choices thoroughly and completely and to select the one that most accurately and precisely answers the question.

13. Restating to Understand

Sometimes, a question on a multiple-choice test is difficult not because of what it asks but because of how it is written. If this is the case, restate the question or answer choice in different words. This process serves a couple of important purposes. First, it forces you to concentrate on the core of the question. In order to rephrase the question accurately, you have to understand it well. Rephrasing the question will concentrate your mind on the key words and ideas. Second, it will present the information to your mind in a fresh way. This process may trigger your memory and render some useful scrap of information picked up while studying.

14. True Statements

Sometimes an answer choice will be true in itself, but it does not answer the question. This is one of the main reasons why it is essential to read the question carefully and completely before proceeding to the answer choices. Too often, test takers skip ahead to the answer choices and look for true statements. Having found one of these, they are content to select it without reference to the question above. Obviously, this provides an easy way for test makers to play tricks. The savvy test taker will always read the entire question before turning to the answer choices. Then, having settled on a correct answer choice, he or she will refer to the original question and ensure that the selected answer is relevant. The mistake of choosing a correct-but-irrelevant answer choice is especially common on questions related to specific pieces of objective knowledge. A prepared test taker will have a wealth of factual knowledge at his or her disposal, and should not be careless in its application.

15. No Patterns

One of the more dangerous ideas that circulates about multiple-choice tests is that the correct answers tend to fall into patterns. These erroneous ideas range from a belief that B and C are the most common right answers, to the idea that an unprepared test-taker should answer "A-B-A-C-A-D-A-B-A." It cannot be emphasized enough that pattern-seeking of this type is exactly the WRONG way to approach a multiple-choice test. To begin with, it is highly unlikely that the test maker will plot the correct answers according to some predetermined pattern. The questions are scrambled and delivered in a random order. Furthermore, even if the test maker was following a pattern in the assignation of correct answers, there is no reason why the test taker would know which pattern he or she was using. Any attempt to discern a pattern in the answer choices is a waste of time and a distraction from the real work of taking the test. A test taker would be much better served by extra preparation before the test than by reliance on a pattern in the answers.

FREE DVD OFFER

Don't forget that doing well on your exam includes both understanding the test content and understanding how to use what you know to do well on the test. We offer a completely FREE Test Taking Tips DVD that covers world class test taking tips that you can use to be even more successful when you are taking your test.

All that we ask is that you email us your feedback about your study guide. To get your **FREE Test Taking Tips DVD**, email freedvd@studyguideteam.com with "FREE DVD" in the subject line and the following information in the body of the email:

- The title of your study guide.
- Your product rating on a scale of 1-5, with 5 being the highest rating.
- Your feedback about the study guide. What did you think of it?
- Your full name and shipping address to send your free DVD.

Introduction to the ISEE Lower Exam

Function of the Test

The Lower Level ISEE (Independent School Entrance Exam) is a test, offered by the Educational Records Bureau (ERB), that is designed to be used for admission assessment at independent schools for entrance to fifth and sixth grades. Two other ISEE exams cover students seeking to enter other grades. Accordingly, the typical test taker is usually a prospective fifth or sixth grade student at a private school in the United States. The test is also used by a few international schools, primarily those catering to American parents.

ISEE scores are available to the test taker and to schools the test taker is seeking admission to. They are typically used only by such schools, and only as part of the admissions process.

Test Administration

The test is available in both computer and paper versions. The computer version can be taken online, allowing it to be administered at any time and date. The test may also be administered at ERB member schools, ERB offices, and any of 400 plus Prometric testing sites.

Upon arrival at the testing site, test takers present a verification letter or identification and get checked in. Test takers are encouraged to ask questions for clarification before the exam begins, as administrators are not permitted to discuss the test questions once testing begins. Test takers are asked to bring four #2 pencils and two pens.

Test takers may register for the Lower Level ISEE no more than three times during a given year, once each in any or all of three testing seasons. The testing seasons are fall (August through November), winter (December through March), and spring/summer (April through July). Reasonable accommodations are available for test takers with documented disabilities under the Americans with Disabilities Act.

Test Format

The content of the Lower Level ISEE is based on standards prepared by organizations including the National Council of Teachers of English, the International Reading Association, and the National Council of Teachers of Mathematics. The test consists of four multiple-choice sections and one essay section. A test taker's ISEE score is based on their performance on the four multiple choice sections. The essay is not graded, but is included with the scores when they are sent to a school. A breakdown of the sections is as follows:

Section	Content	Questions	Time
Verbal Reasoning	Multiple choice, scored	34	20
Quantitative Reasoning	Multiple choice, scored	38	35
Reading Comprehension	Multiple choice, scored	25	25
Mathematics Achievement	Multiple choice, scored	30	30
Essay	Written, unscored	NA	30

Scoring

Scores are based only on the number of correct answers provided. There is no penalty for guessing incorrectly, aside from the missed opportunity to provide another correct answer. That total number of correct answers becomes a raw score, which is then converted to a scaled score between 760 and 940. There is no set passing score on the exam. Instead, scores are reviewed by schools in conjunction with other factors in determining admissions decisions.

Scores are first provided to the test taker's family. The family may then decide whether to release the report to schools, and to which schools to release it. Scores may be received as soon as a couple days after completion of an exam.

Recent/Future Developments

ERB recently set the limit on retakes at one per testing season. It also instituted a rule allowing test takers' families to review scores before a school does.

Practice Test #1

Verbal Reasoning

Synonyms

Each of the questions below has one word. The one word is followed by four words or phrases. Please select one answer whose meaning is closest to the word in capital letters.

1. COINCIDE:
   a. acquiesce
   b. deceive
   c. marvel
   d. quench

2. LOFTY:
   a. deft
   b. elevated
   c. frigid
   d. innate

3. BRAWL:
   a. boycott
   b. engross
   c. fight
   d. fuse

4. DWELL:
   a. accompany
   b. bluster
   c. compel
   d. inhabit

5. ANONYMOUS:
   a. aroma
   b. conspicuous
   c. flammable
   d. nameless

6. GORGE:
   a. hoard
   b. hoax
   c. infest
   d. overeat

7. ABSURD:
   a. benevolent
   b. cordial
   c. mediocre
   d. ridiculous

8. HECTIC:
   a. chaotic
   b. mellow
   c. peculiar
   d. ravenous

9. DELICATE:
   a. amiable
   b. fragile
   c. malicious
   d. nonchalant

10. SUBDUED:
   a. animated
   b. diligent
   c. fickle
   d. quiet

11. ENGAGING:
   a. charming
   b. indifferent
   c. sentimental
   d. steadfast

12. HEADSTRONG:
   a. meek
   b. robust
   c. stubborn
   d. vivacious

13. FRET:
   a. emulate
   b. flatter
   c. toil
   d. worry

14. FATIGUE:
   a. fortitude
   b. struggle
   c. vigor
   d. weariness

15. ULTIMATE
    a. final
    b. indictment
    c. maxim
    d. tragedy

16. ANXIETY
    a. bitterness
    b. harmony
    c. strife
    d. worry

17. FEEBLE
    a. cynical
    b. docile
    c. frail
    d. resolute

Sentence Completion

Select the word or phrase that most correctly completes the sentence.

18. At one time, the Roman Empire was one of the most __________ military, economic, political, and cultural forces in the world.
    a. demure
    b. disappointing
    c. lax
    d. robust

19. Since the tadpole had a __________ outer covering, the biologist could not tell what the internal organs looked like.
    a. bright
    b. clear
    c. sheer
    d. solid

20. The students were saying nice things about their friend Michael that __________ his reputation.
    a. boosted
    b. disgraced
    c. erased
    d. removed

21. The boy looked like he had just risen from bed judging by his __________ appearance.
    a. jaded
    b. lively
    c. slovenly
    d. subdued

22. Even after eating three full meals and several snacks, the athlete's hunger was not __________.
    a. affordable
    b. broken
    c. perilous
    d. satisfied

23. Alison Creek, a waterway located in California, could not be used as a source of drinking water because of its __________ quality.
    a. admirable
    b. beneficial
    c. detrimental
    d. unique

24. Josephine did not have much time to complete her test, so she started concentrating on the main topics instead of on the __________ material.
    a. critical
    b. important
    c. irrelevant
    d. necessary

25. Juan is a talented painter best known for his use of __________.
    a. color
    b. hyperbole
    c. onomatopoeia
    d. simile

26. Chrissy is a great team leader because she is able to __________ her team to produce rapid results.
    a. confuse
    b. daunt
    c. disable
    d. inspire

27. The students were so exhausted after the week of practice testing that they acted __________ when asked to perform one more test.
    a. animated
    b. energized
    c. motivated
    d. sluggish

28. The music tonight was so __________ that I had to cover my ears to dilute the sound.
    a. muffled
    b. placid
    c. rowdy
    d. serene

29. The thief __________ with the purse before anyone knew what had happened.
    a. bolted
    b. lagged
    c. slowed
    d. waned

30. The teacher recognized the immature writing style of the first-grader because his papers used __________ language.
   a. excellent
   b. mediocre
   c. perfect
   d. prestige

31. Much to the __________ of the unprepared students, the teacher called on students randomly to deliver unscripted speeches.
   a. assurance
   b. dismay
   c. elation
   d. pleasure

32. Margaret moved into a new house that was __________ and extravagant.
   a. destitute
   b. fancy
   c. puny
   d. scanty

33. Rachel wanted __________ for the time she had spent going through orientation.
   a. accumulation
   b. compensation
   c. dedication
   d. escalation

34. The __________ of the river was astounding; the group could see all the way down to the bottom.
   a. clarity
   b. murkiness
   c. shadow
   d. gloom

Quantitative Reasoning

1. Express as an improper fraction $8\frac{3}{7}$.
   a. $\frac{11}{7}$
   b. $\frac{5}{3}$
   c. $\frac{21}{8}$
   d. $\frac{59}{7}$

2. Which of the following expressions is equivalent to $\frac{8}{7}$?

a. $\frac{7}{8} + 7$

b. $\frac{1}{7} + \frac{8}{1}$

c. $\frac{1}{8} + \frac{1}{8} + \frac{1}{8} + \frac{1}{8} + \frac{1}{8} + \frac{1}{8} + \frac{1}{8}$

d. $\frac{1}{7} + \frac{1}{7} + \frac{1}{7} + \frac{1}{7} + \frac{1}{7} + \frac{1}{7} + \frac{1}{7} + \frac{1}{7}$

3. Which of the following equations best represents the problem below?

The width of a rectangle is 2 centimeters less than the length. If the perimeter of the rectangle is 44 centimeters, then what are the dimensions of the rectangle?

a. $l \times (l - 2) = 44$
b. $2l + 2(l - 2) = 44$
c. $l + 2) + (l + 2) + l = 48$
d. $(l + 2) + (l + 2) + l = 44$

4. Chris walks $\frac{4}{7}$ of a mile to school and Tina walks $\frac{5}{9}$ of a mile. Which student covers more distance on the walk to school?

a. Chris, because $\frac{4}{7} > \frac{5}{9}$

b. Chris, because $\frac{4}{7} < \frac{5}{9}$

c. Tina, because $\frac{5}{9} > \frac{4}{7}$

d. Tina, because $\frac{5}{9} < \frac{4}{7}$

5. Express as a reduced mixed number $\frac{54}{15}$.

a. $3\frac{1}{54}$

b. $3\frac{3}{54}$

c. $3\frac{1}{15}$

d. $3\frac{3}{5}$

6. Kareem arrived for his 9:00 a.m. appointment with the dentist 15 minutes early. He was taken back to see the see the dentist 30 minutes after he arrived. His cleaning took 45 minutes once he was taken back. What time was he done with the dental cleaning?

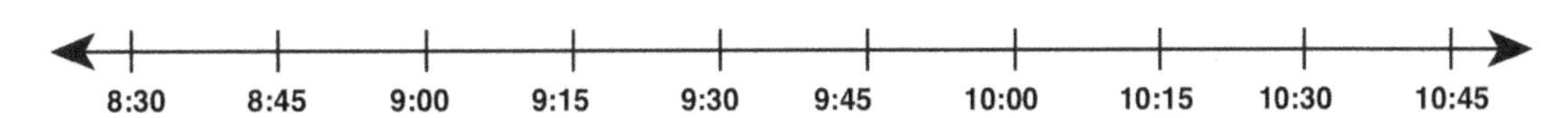

a. 9:45 a.m.
b. 10:00 a.m.
c. 10:15 a.m.
d. 10:30 a.m.

7. Arrange the following numbers from least to greatest value:

$0.85, \frac{4}{5}, \frac{2}{3}, \frac{91}{100}$

a. $0.85, \frac{4}{5}, \frac{2}{3}, \frac{91}{100}$

b. $\frac{4}{5}, 0.85, \frac{91}{100}, \frac{2}{3}$

c. $\frac{2}{3}, \frac{4}{5}, 0.85, \frac{91}{100}$

d. $0.85, \frac{91}{100}, \frac{4}{5}, \frac{2}{3}$

8. Keith's bakery had 252 customers go through its doors last week. This week, that number increased to 378. Express this increase as a percentage.

a. 12%
b. 26%
c. 35%
d. 50%

9. The following graph compares the various test scores of the top three students in each of these teacher's classes. Based on the graph, which teacher's students had the lowest range of test scores?

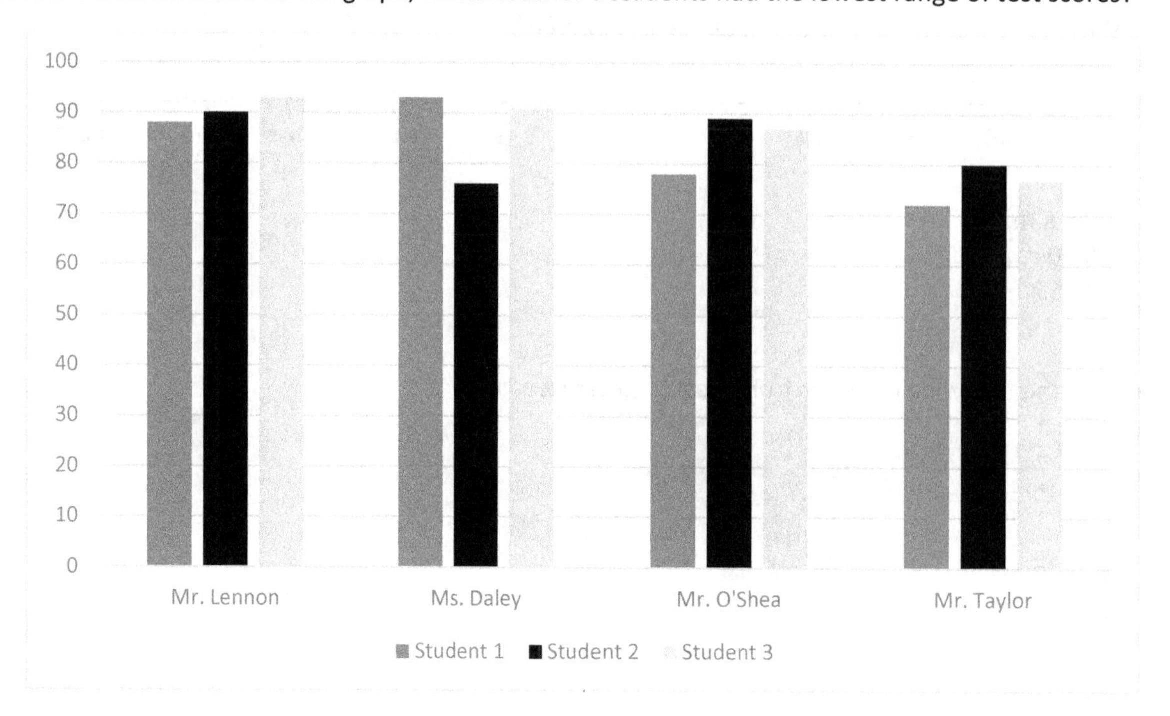

a. Ms. Daley
b. Mr. Lennon
c. Mr. O'Shea
d. Mr. Taylor

10. Based on the graph, how many more points did student 3 get on her test in Mr. O'Shea's class than on her test in Mr. Taylor's class?

a. 5
b. 8
c. 10
d. 14

11. Bernard can make $80 per day. If he needs to make $300 and only works full days, how many days will this take?

a. 2
b. 3
c. 4
d. 5

12. Use the picture graph below to answer the question.

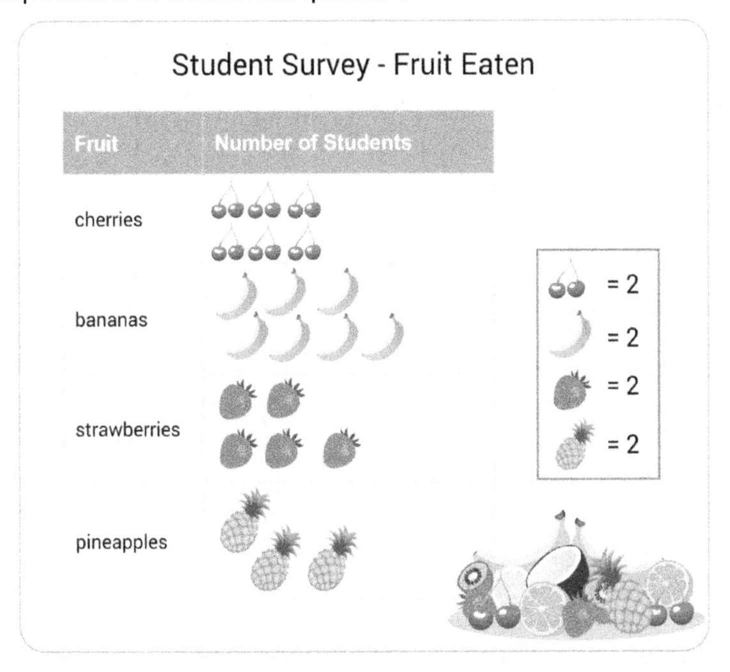

How many more bananas were eaten than pineapples?

a. 4
b. 5
c. 7
d. 8

13. When rounding 423.2769 to the nearest hundredth, which place value would be used to decide whether to round up or round down?

a. Ten-thousandth
b. Thousandth
c. Hundredth
d. Thousand

14. Which of the following are correct labels for the chart below?

Input	Calculation (Input × 3)	Output
1	1 × 3	3
2	2 × 3	6
3	3 × 3	9
4	4 × 3	12

a. Input: number of chairs
   Calculation: number of chairs × number of legs on a chair
   Output: number of rubber feet for chairs to order
b. Input: number of wheels on a tricycle
   Calculation: number of tricycles
   Output: number of wheels in inventory
c. Input: number of tricycles
   Calculation: number of wheels on a tricycle
   Output: number of wheels in inventory
d. Input: number of booties for dogs
   Calculation: number of dogs
   Output: number of booties in inventory

15. Five students take a test. The scores of the first four students are 80, 85, 75, and 60. If the median score is 80, which of the following could NOT be the score of the fifth student?
a. 60
b. 80
c. 85
d. 100

16. Ten students take a test. Five students get a 50. Four students get a 70. If the average score is 55, what was the last student's score?
a. 20
b. 40
c. 50
d. 60

17. A company invests $50,000 in a building where they can produce saws. If the cost of producing one saw is $40, then which expresses the amount of money the company pays? The variable *x* is the number of saws produced.

a. $50{,}000x + 40$
b. $40x - 50{,}000$
c. $40x + 50{,}000$
d. $x - 50{,}000 + 40$

18. A six-sided die is rolled. What is the probability that the roll is 1 or 2?

a. $\frac{1}{6}$
b. $\frac{1}{4}$
c. $\frac{1}{3}$
d. $\frac{1}{2}$

19. A student gets an 85% on a test with 20 questions. How many answers did the student solve correctly?

a. 15
b. 16
c. 17
d. 18

The table below shows the number of students in Ms. Jackson' class who play each sport.

Sports Played By Students in Ms. Jackson's Class

Sport	Frequency
Soccer	~~IIII~~ II
Swimming	I
Track	III
Baseball	~~IIII~~ I
Basketball	~~IIII~~ I
Tennis	II

20. Which of the following dot plots correctly represents the data in the table on the previous page?

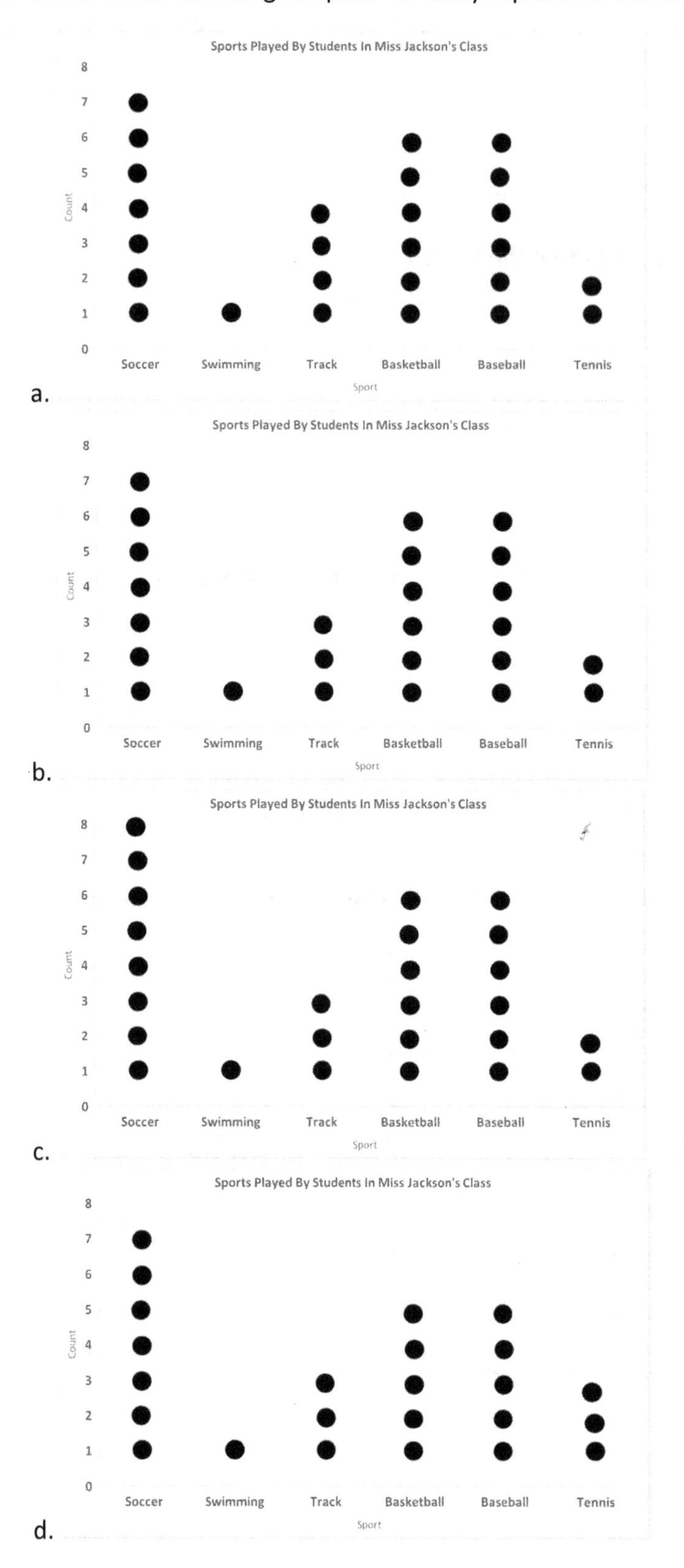

21. Jessica buys 10 cans of paint. Red paint costs $1 per can and blue paint costs $2 per can. In total, she spends $16. How many red cans did she buy?
    a. 2
    b. 3
    c. 4
    d. 5

22. Which item taught in the classroom would allow students to correctly find the solution to the following problem: A clock reads 5:00 am. What is the measure of the angle formed by the two hands of that clock?
    a. Two adjacent angles sum up to 180 degrees.
    b. Two complementary angles sum up to 180 degrees.
    c. Each time increment on an analog clock measures 30 degrees.
    d. Each time increment on an analog clock measures 90 degrees.

23. How is the number -4 classified?
    a. Real, irrational
    b. Real, rational, integer
    c. Real, rational, integer, natural
    d. Real, rational, integer, whole, natural

24. A solution needs 5 mL of saline for every 8 mL of medicine given. How much saline is needed for 45 mL of medicine?
    a. $\frac{45}{8}$ mL
    b. $\frac{225}{8}$ mL
    c. 28 mL
    d. 72 mL

25. What is the solution to $7 \times 7 \div 7 + 7 - 7 \div 7$?
    a. 0
    b. 13
    c. 28
    d. 49

26. Which of the following could be used to show $\frac{3}{7} < \frac{5}{6}$ is a true statement?
    a. A bar graph
    b. A number line
    c. An area model
    d. Base 10 blocks

27. Which of the following shows a line of symmetry?

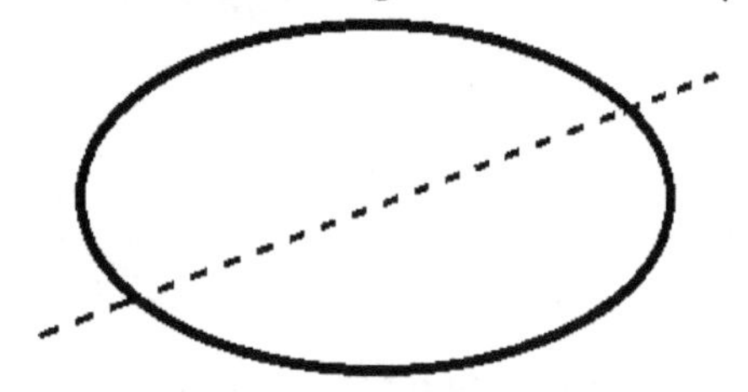

a.

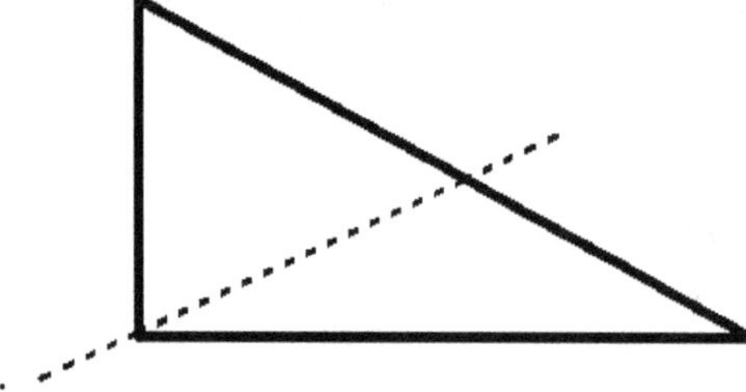

b.

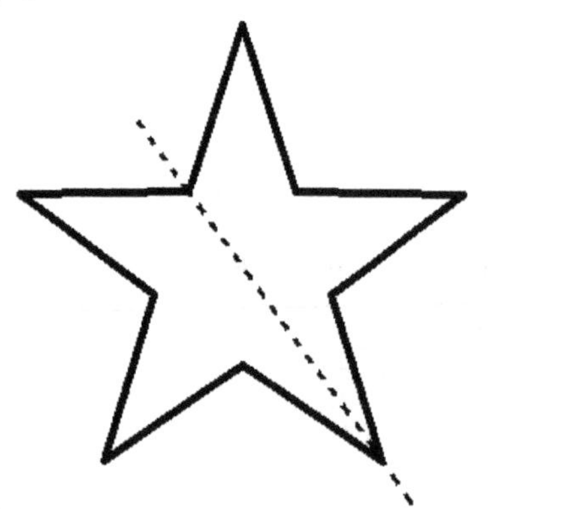

c.

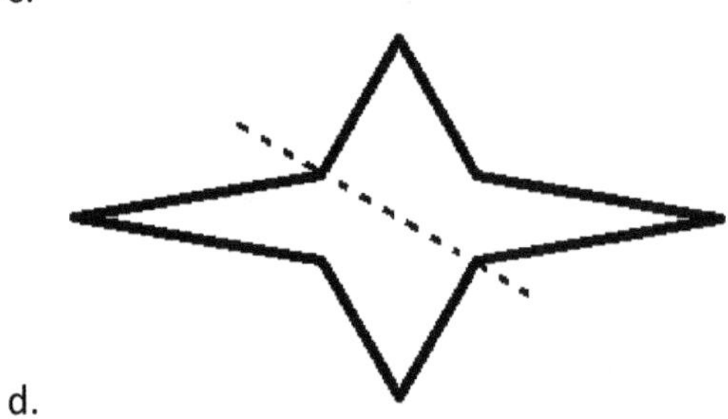

d.

28. A ball is drawn at random from a ball pit containing 8 red balls, 7 yellow balls, 6 green balls, and 5 purple balls. What's the probability that the ball drawn is yellow?

a. $\frac{1}{26}$

b. $\frac{7}{26}$

c. $\frac{19}{26}$

d. 1

29. What would the equation be for the following problem?

3 times the sum of a number and 7 is greater than or equal to 32

a. $3n + 7 = 32$
b. $3n + 7 \leq 32$
c. $3(n + 7) = 32$
d. $3(n + 7) \geq 32$

30. An accounting firm charted its income on the following pie graph. If the total income for the year was $500,000, how much of the income was received from Audit and Taxation Services?

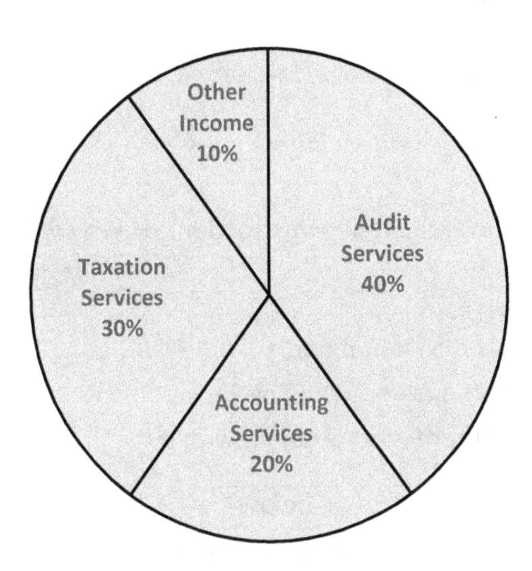

a. $150,000
b. $200,000
c. $300,000
d. $350,000

31. Which of the following represents one hundred eighty-two million, thirty-six thousand, four hundred twenty-one and three hundred fifty-six thousandths?

a. 182,036,421.0356
b. 182,036,421.356
c. 182,000,036,421.0356
d. 182,000,036,421.356

32. 32 is 25% of what number?

a. 8
b. 12.65
c. 64
d. 128

33. Which of the following is a mixed number?

a. $\frac{1}{4}$

b. $\frac{16}{3}$

c. 16

d. $16\frac{1}{2}$

34. The phone bill is calculated each month using the equation $c = 50g + 75$. The cost of the phone bill per month is represented by c, and g represents the gigabytes of data used that month. What is the value and interpretation of the slope of this equation?

a. 75 dollars per day
b. 50 dollars per day
c. 75 gigabytes per day
d. 50 dollars per gigabyte

35. A student answers a problem with the following fraction: $\frac{3}{15}$. Why would this be considered incorrect?

a. It is not expressed in decimal form.
b. It needs to be converted to a mixed number.
c. It is not simplified. The correct answer would be $\frac{1}{5}$.
d. It is in the correct form, and there is no problem with it.

36. Which inequality represents the number line below?

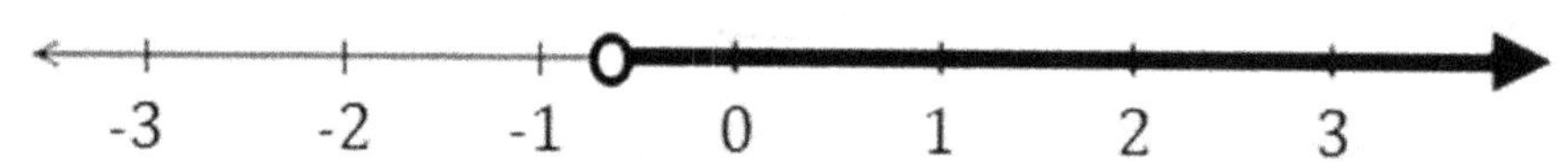

a. $4x + 5 < 8$
b. $-4x + 5 < 8$
c. $-4x + 5 > 8$
d. $4x - 5 > 8$

37. Which of the following statements is true about the two lines below?

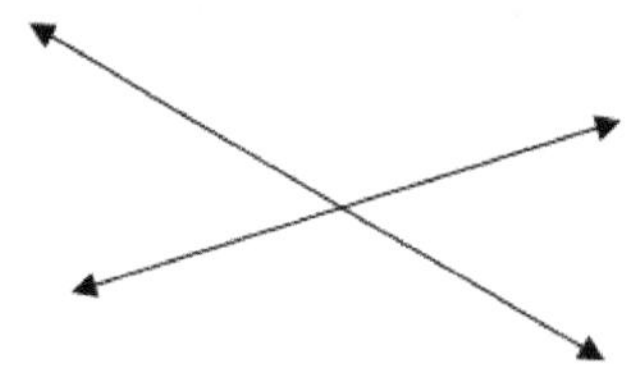

a. The two lines are parallel but not perpendicular.
b. The two lines are perpendicular but not parallel.
c. The two lines are both parallel and perpendicular.
d. The two lines are neither parallel nor perpendicular.

38. For a group of 20 men, the median weight is 180 pounds and the range is 30 pounds. If each man gains 10 pounds, which of the following would be true?
    a. The median weight will increase, and the range will remain the same.
    b. The median weight and range will both remain the same.
    c. The median weight will stay the same, and the range will increase.
    d. The median weight and range will both increase.

Reading Comprehension

Read the following passage, and then answer questions 1–5.

1 Do you want to vacation at a Caribbean island destination? Who wouldn't want a tropical
2 vacation? Visit one of the many Caribbean islands where visitors can swim in crystal blue waters,
3 swim with dolphins, or enjoy family-friendly resorts and activities. Every island offers a unique
4 and dazzling vacation destination. Choose from these islands: Aruba, St. Lucia, Barbados,
5 Anguilla, St. John, and so many more. A Caribbean island destination will be the best and most
6 refreshing vacation ever … no regrets!

1. What is the topic of the passage?
    a. Resorts
    b. Activities
    c. Tropical vacation
    d. Caribbean island destinations

2. What is/are the supporting detail(s) of this passage?
    a. Beaches
    b. Local events
    c. Family activities
    d. Cruising to the Caribbean

Read the following sentence, and answer the question below.

"A Caribbean island destination will be the best and most refreshing vacation ever … no regrets!"

3. What is this sentence an example of?
    a. Device
    b. Fact
    c. Fallacy
    d. Opinion

4. What is the author's purpose of this passage?
    a. Entertain readers
    b. Persuade readers
    c. Inform or teach readers
    d. Share a moral lesson to readers

5. What would you most likely bring to a Caribbean island vacation?
   a. Football helmet
   b. Hiking boots and tent
   c. Ski jacket and long pants
   d. Swimsuit or swim trucks and goggles

Read the following passage, and then answer questions 6–10.

1 Even though the rain can put a damper on the day, it can be helpful and fun, too. For one, the
2 rain helps plants grow. Without rain, grass, flowers, and trees would be deprived of vital
3 nutrients they need to develop. Not only does the rain help plants grow, but on days where
4 there are brief spurts of sunshine, rainbows can appear. The rain reflects and refracts the light,
5 creating beautiful rainbows in the sky. Finally, puddle jumping is another fun activity that can be
6 done in or after the rain. Therefore, the rain can be helpful and fun.

6. What is the *cause* in this passage?
   a. Rain
   b. Rainbows
   c. Plants growing
   d. Puddle jumping

Read the following sentence, and answer the question below.

"Without rain, grass, flowers, and trees would be deprived of vital nutrients they need to develop."

7. In this sentence, the author is using what literary device regarding the grass, flowers, and trees?
   a. Comparing
   b. Contrasting
   c. Describing
   d. Transitioning

8. In the same sentence, what is most likely the meaning of *vital?*
   a. Dangerous
   b. Energetic
   c. Necessary
   d. Truthful

9. What is an *effect* in this passage?
   a. Rain
   b. Rainbows
   c. Weather
   d. Brief spurts of sunshine

10. Which of the following is a reason why rain is helpful and fun?
   a. Rain can put a damper on the day.
   b. Plants are deprived of vital nutrients.
   c. Rain prevents rainbows from appearing.
   d. Puddle jumping can be done in or after the rain.

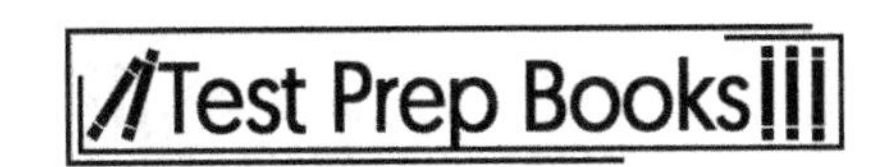

Read the following passage, and then answer questions 11–15.

1 Meet Lola. Lola is an overly friendly Siberian husky who loves her long walks, digs holes for days,
2 and sheds unbelievably . . . like a typical Siberian husky. Lola has to be brushed and brushed and
3 brushed—did I mention that she has to be brushed all the time! On her long walks, Lola loves
4 making friends with new dogs and kids. A robber could break into our house, and even though
5 they may be intimidated by Lola's wolf-like appearance, the robber would be shocked to learn
6 that Lola would most likely greet them with kisses and a tail wag. She makes friends with
7 everyone! Out of all the dogs we've ever owned, Lola is certainly one of a kind in many ways.

11. Based on the passage, what does the author imply?
    a. Siberian huskies are good guard dogs.
    b. Siberian huskies are easy to take care of.
    c. Siberian huskies should not be around children.
    d. Siberian huskies are great pets but require a lot of time and energy.

12. What word best describes the author of this passage because of their own experience with Siberian huskies?
    a. Biased
    b. Hasty
    c. Impartial
    d. Irrational

13. Based on the passage, we can infer what about Lola's owner (the narrator)?
    a. It is a man.
    b. It is a woman.
    c. It is a new dog owner.
    d. It is an experienced dog owner.

14. Which of the following best describes Lola?
    a. Aggressive
    b. Amiable
    c. Anxious
    d. Apprehensive

15. Based on the information in the passage, which of the following dogs most likely looks like Lola?

a.   b.   c.   d.

Read the following passage, and then answer questions 16–20.

1 Learning how to write a ten-minute play may seem like a challenging task at first; but, if you
2 follow a simple creative writing strategy, similar to writing a story, you will be able to write a
3 successful drama. The first step is to open your story as if it is a puzzle to be solved. This will
4 allow the reader to engage with the story and to solve the story with you, the author.
5 Immediately provide descriptive details that set the main idea, the tone, and the mood
6 according to the theme you have in mind. Next, use dialogue to reveal the attitudes and
7 personalities of each of the characters who have a key part in the unfolding story. Show images
8 on stage to speed up the dialogue; remember, one picture speaks a thousand words. As the play
9 progresses, the protagonist must cross the point of no return in some way; this is the climax of
10 the story. Then, as in a written story, you create a resolution to the life-changing event of the
11 protagonist.

16. Based on the passage above, select the statement that is true.
    a. Writing a ten-minute play is very difficult.
    b. Providing descriptive details is not necessary.
    c. The climax of the story sets the theme you have in mind.
    d. Descriptive details give clues to the play's intended mood and tone.

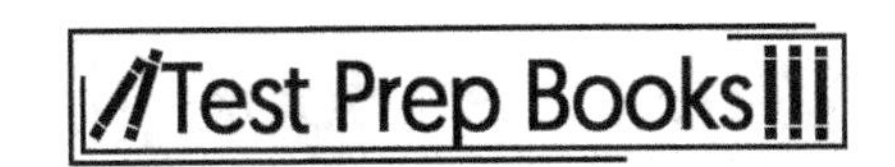

17. Which of the following is the most likely meaning for the phrase "one picture speaks a thousand words" in the following sentence?

"Show images on stage to speed up the dialogue; remember, one picture speaks a thousand words."

a. Pictures can tell stories as well as, if not better than, words.
b. Playwrights should be sure to add videos to speed up the dialogue.
c. Audio-video technology should be used to enhance scenery in a play.
d. Playwrights should include an image after every 1000 words of dialogue.

18. What is the meaning of the word *protagonist?*
a. Actor
b. Student
c. Playwright
d. Main character

19. In the passage above, the writer suggests that writing a ten-minute play is doable for a new playwright for which of the following reasons?
a. Dialogue is not necessary if you have images.
b. The format follows similar strategies of writing a narrative story.
c. It took the author of the passage only one week to write their first play.
d. There are no particular themes or points to unravel in a ten-minute play.

20. What is the meaning of the word *resolution*?
a. Cause
b. Conclusion
c. Conversation
d. Courage

Read the following passage, and then answer questions 21–25.

1 Overall, we won the Little League championship game! Max hit a winning home run, and we all
2 cheered as he rounded home plate. It was an astonishing win because the other team wins
3 every year and we were down the whole game until the final inning. Our team hoisted the
4 championship trophy up into the air and celebrated with joy. It was such a great game. After the
5 game, my coach took my whole team to the diner and we got burgers, fries, and chocolate
6 milkshakes. Max got grilled cheese because he is a vegetarian. This was the first championship
7 game that our team has won in twenty years. My coach gave a speech while we were eating and
8 said he was proud of our perseverance.

21. What is mostly likely the meaning of *astonishing* in the following sentence?

"It was an astonishing win because the other team wins every year and we were down the whole game until the final inning."

a. Celebrated
b. Close
c. Expected
d. Surprising

22. What is the main topic of the passage?
a. A basketball team's victory
b. A meal at the diner with his team
c. Vegetarian food options at a diner
d. Winning the baseball championship

23. Which of the following is the best description of the tone or mood of the passage?
a. Disappointed
b. Excited
c. Informational
d. Nervous

24. Which of the following can readers infer about Max?
a. He is overweight.
b. He does not eat bacon.
c. He is the narrator of the passage.
d. He has been on the team for twenty years.

25. What is the meaning of the word *perseverance*?
a. Decision
b. Dedication
c. Laziness
d. Weakness

Math Achievement

1. If Amanda can eat two times as many mini cupcakes as Marty, what would the missing values be for the following input-output table?

Input (number of cupcakes eaten by Marty)	Output (number of cupcakes eaten by Amanda)
1	2
3	
5	10
7	
9	18

a. 3, 11
b. 4, 12
c. 6, 10
d. 6, 14

2. Last month you made $20 from babysitting, $35 from a lemonade stand, $15 from weeding a neighbor's yard, and earned $25 in allowance. You also purchased a video game for $50, went to a movie for $12, and bought $15 worth of snacks. After balancing your budget, were you able to cover your expenses last month with your earnings?

a. No. You spent more than you earned.
b. Yes. You did not spend any money last month.
c. No. You broke even and spent exactly as much as you earned.
d. Yes. The amount of money you earned exceeded your expenses.

3. If $6t + 4 = 16$, what is t?

a. 1
b. 2
c. 3
d. 4

4. Divide, express with a remainder $1{,}202 \div 44$.

a. $2\frac{7}{22}$
b. $7\frac{2}{7}$
c. $27\frac{2}{7}$
d. $27\frac{7}{22}$

5. 1 kilometer is how many centimeters?
   a. 100 cm
   b. 1,000 cm
   c. 10,000 cm
   d. 100,000 cm

6. Add 78.654 and 4.900. Round to the nearest hundredth.
   a. 82.4
   b. 83.55
   c. 86.75
   d. 91.351

7. Change $3\frac{3}{5}$ to a decimal.
   a. 0.28
   b. 3.6
   c. 4.67
   d. 5.3

8. Add $103{,}678 + 487$
   a. 103,191
   b. 103,550
   c. 104,165
   d. 104,265

9. Solve this equation:

$$9x + x - 7 = 16 + 2x$$

   a. $x =$ -4
   b. $x = \frac{9}{8}$
   c. $x = \frac{23}{8}$
   d. $x = 3$

10. A bucket can hold 11.4 liters of water. A kiddie pool needs 35 gallons of water to be full. How many times will the bucket need to be filled to fill the kiddie pool? ($1\ gallon = 3.8\ liters$)
   a. 11
   b. 12
   c. 35
   d. 45

11. The soccer team is selling donuts to raise money to buy new uniforms. For every box of donuts they sell, the team receives $3 towards their new uniforms. There are 15 people on the team. How many boxes does each player need to sell in order to raise $270 for their new uniforms?
  a. 5
  b. 6
  c. 7
  d. 9

12. Subtract $701.1 - 52.33$.
  a. 638.43
  b. 648.77
  c. 652.77
  d. 753.43

The following stem-and-leaf plot shows plant growth in cm for a group of tomato plants.

Stem	Leaf
2	0 2 3 6 8 8 9
3	2 6 7 7
4	7 9
5	4 6 9

13. What is the range of measurements for the tomato plants' growth?
  a. 29 cm
  b. 37 cm
  c. 39 cm
  d. 59 cm

14. How many plants grew more than 35 cm?
  a. 4 plants
  b. 5 plants
  c. 8 plants
  d. 9 plants

15. Add and express in reduced form $\frac{14}{33} + \frac{10}{11}$.
   a. $\frac{2}{11}$
   b. $\frac{6}{11}$
   c. $\frac{4}{3}$
   d. $\frac{44}{33}$

16. The ratio of boys to girls in Mrs. Lair's kindergarten class is 3 to 2. If there are 30 kids in the class, how many of them are boys?
   a. 3
   b. 9
   c. 12
   d. 18

17. If Sarah reads at an average rate of 21 pages in four nights, how long will it take her to read 140 pages?
   a. 6 nights
   b. 8 nights
   c. 26 nights
   d. 27 nights

18. Subtract $112{,}076 - 1{,}243$.
   a. 110,319
   b. 110,833
   c. 113,319
   d. 113,833

19. If a school has 550 boys and 635 girls, what is the percentage of the students that are girls?
   a. 46.4
   b. 50.6
   c. 53.6
   d. 86.6

20. What are the coordinates of the two points marked with dots on this coordinate plane?

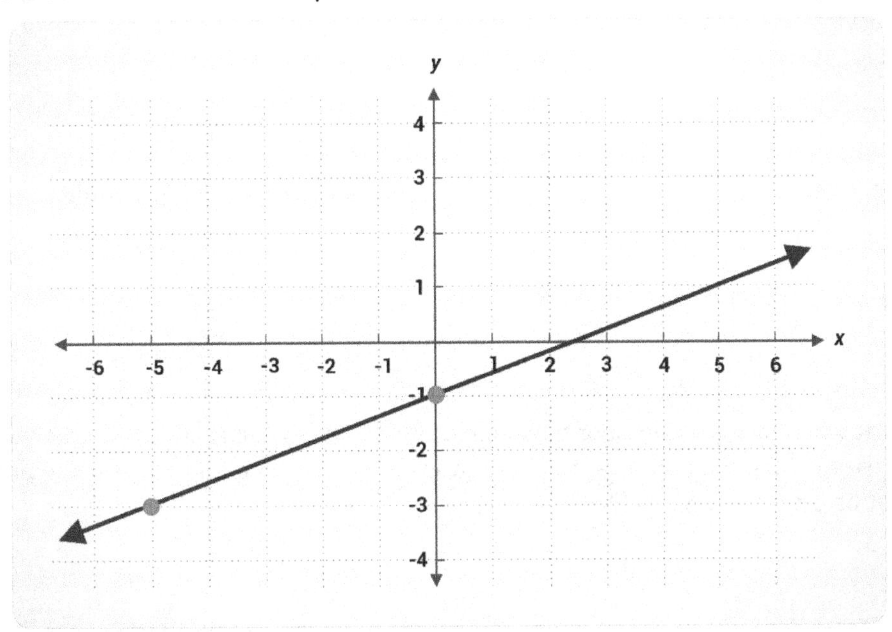

a. (-3, -5) and (-1, 0)
b. (5, 3) and (0, 1)
c. (-5, -3) and (0, -1)
d. (-3, -5) and (0, -1)

21. $864 \div 36 =$
a. 18
b. 24
c. 25
d. 34

22. Last month you made $45 from lawnmowing, $20 from tutoring a friend, and earned $75 from your paper route. Last month, you also spent $10 on a graphic novel, $85 on whitewater rafting, and $50 on admission to an amusement park. After balancing your budget, were you able to cover your expenses last month with your earnings?
a. Yes. You did not spend any money last month.
b. No. You broke even and spent exactly as much as you earned.
c. Yes. The amount of money you earned exceeded your expenses.
d. No. You spent more than you earned and need to borrow money from your savings account to cover your expenses.

23. Twenty is 40 percent of what number?
   a. 8
   b. 50
   c. 200
   d. 500

24. Kyle's average test grade in math was an 85. His 4 test grades were 87, 85, 78, and x. What is the value of x?
   a. 83.75
   b. 85
   c. 85.6
   d. 90

25. Nina has a jar where she puts her loose change at the end of each day. There are 13 quarters, 25 dimes, 18 nickels, and 30 pennies in the jar. If she chooses a coin at random, what is the probability that the coin will not be a penny or a dime?
   a. 0.34
   b. 0.36
   c. 0.56
   d. 0.64

26. Two cards are drawn from a shuffled deck of 52 cards. What's the probability that both cards are Kings if the first card isn't replaced after it's drawn?
   a. $\frac{1}{221}$
   b. $\frac{1}{169}$
   c. $\frac{1}{13}$
   d. $\frac{4}{13}$

27. Write the expression for six less than three times the sum of twice a number and one.
   a. $2x + 1 - 6$
   b. $3x + 1 - 6$
   c. $3(x + 1) - 6$
   d. $3(2x + 1) - 6$

28. If $3x - 4 + 5x = 8 - 10x$, what is the value of x?
   a. -6
   b. 0.5
   c. 0.67
   d. 6

29. A rectangle has a length that is 5 feet longer than three times its width. If the perimeter is 90 feet, what is the length in feet?
   a. 10
   b. 20
   c. 25
   d. 35

30. Johnny is a competitive runner. He knows that a 5k run is 3.1 miles. About how many kilometers does he need to run if he wants to run roughly 15 miles?

  a. 9 km
  b. 18 km
  c. 21 km
  d. 25 km

Essay

Select a topic from the list below and write an essay. You may organize your essay on another sheet of paper.

Topic 1: If you could choose one hobby to do for the rest of your life, what would it be and why?

Topic 2: What is the best advice you could give someone at this moment?

Topic 3: If you could own any business, what would it be and why?

Answer Explanations #1

Verbal Reasoning

Synonyms

1. A: The words *coincide* and *acquiesce* both mean to go along with or agree to something. Choice *B*, deceive, means to be dishonest or betray. Choice *C*, marvel, means to be amazed at something. Choice *D*, quench, means to satisfy a thirst.

2. B: The words *lofty* and *elevated* are the most closely related because they both mean high, soaring, or stately. Choice *B*, deft, means clever or agile. Choice *C*, frigid, means cold in personality or in a physical sense. Choice *D*, innate, means something inherited.

3. C: To *brawl* means to *fight*. Choice *A*, *boycott*, means to refrain from using a good or service. Choice *B*, engross, means to captivate or bewitch. Choice *D*, fuse, means to blend or combine.

4. D: To *dwell* means to *inhabit* or live in. Choice *A*, accompany, means to attend or escort. Choice *B*, bluster, means to bully or intimidate. Choice *C*, compel, means forced to act in a certain way.

5. D: *Anonymous* and *nameless* are synonyms. Choice *A*, aroma, indicates a distinctive smell. Choice *B*, conspicuous, means obvious or apparent. Choice *C*, flammable, means easily set afire.

6. D: To *gorge* means to *overeat*. Choice *A*, hoard, means to accumulate or stockpile. Choice *B*, hoax, means to trick. Choice *C*, to infest, means to invade as a mass quantity.

7. D: The word *absurd* is most closely related to the word *ridiculous*. Choice *A*, benevolent, means good. Choice *B*, cordial, means sociable and friendly. Choice *C*, mediocre, means average or commonplace.

8. A: *Hectic* and *chaotic* are synonyms and both mean disordered, confused, and turbulent. Choice *B*, mellow, means easygoing and is the opposite of hectic. Choice *C*, peculiar, means unusual or strange. Choice *D*, ravenous, means extremely hungry.

9. B: *Delicate* and *fragile* are synonyms. Choice *A*, amiable, means friendly. Choice *C*, malicious, means hateful. Choice *D*, nonchalant, means aloof.

10. D: *Subdued* is most closely related to the word *quiet*. Choice *A*, animated, means lively or full of life. Choice *B*, diligent, means hard-working. Choice *C*, fickle, means not reliable or dependable.

11. A: *Engaging* means to be *charming*. Choice *B*, indifferent, means showing no interest or concern. Choice *C*, sentimental, means over emotional, which isn't quite the same as just being simply charming. Choice *D*, steadfast, means to be loyal to something or someone.

12. C: *Headstrong* means stubborn. Choice *A* meek, means tame or humble. Choice *B*, robust, means strong or sturdy. Choice *D*, vivacious, means lively or spirited.

13. D: To *fret* means to worry. Choice *A*, emulate, means to copy or imitate. Choice *B*, flatter, means to praise excessively. Choice *C*, toil, means to work hard.

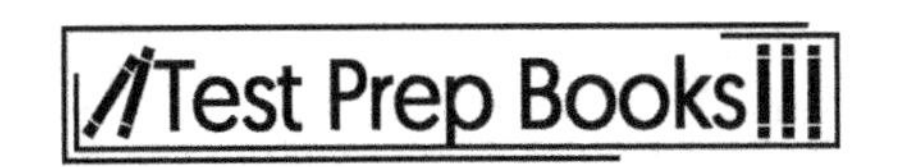

14. D: *Fatigue* means to be very tired or weary. Choice *A*, fortitude, means possessing strength or bravery. Choice *B*, struggle, has several definitions but generally means to have difficulty handling something. Choice *C*, vigor, is an antonym of fatigue and has to do with being lively and full of spirit.

15. A: *Ultimate* is synonyms with *final* or *last*. Choice *B*, indictment, is an accusation. Choice *C*, maxim, is a common proverb or saying. Choice *D*, tragedy, denotes a disaster or adversity in someone's life.

16. D: Having *anxiety* means to have worry or nervousness. Choice *A*, bitterness, isn't quite the same as anxiety, as it means having hostility or anguish. Choice *B*, harmony, means to have balance or agreement, either in music or in a social relationship. Choice *C*, strife, means conflict.

17. C: *Feeble* is most closely related to the word *frail.* Choice *A*, cynical, means to be doubtful or derisive. Choice *B*, docile, means to be submissive. Choice *D*, resolute, means determined or adamant.

Sentence Completion

18. D: *Robust* is the best word here because it means strong and healthy. A "cultural force" would be one with strong military, economic, and political power.

19. D: The word *solid* is the best fit here. If something has a solid outer surface, it would be impossible to view the inner organs.

20. A: We can assume that the students were saying nice things about their friend and that it *boosted* the friend's reputation. The other words mean being dragged down, so they are incorrect.

21. C: *Slovenly* means someone who is untidy or messy in their appearance, so this is the correct answer choice. The other answer choices do not fit the context of the sentence.

22. D: The athlete's hunger was not *satisfied* even after eating three full meals and several snacks. Broken, perilous, and affordable are not characteristics of hunger, so these are incorrect.

23. C: The word *detrimental* means damaging or unhealthy. Therefore, if the water cannot be used as a source to drink from, it must be unhealthy, or *detrimental,* to someone's health.

24. C: We can rule Choices *A*, *B*, and *D* out. Critical, important, and necessary, all mean that the material is valuable to Josephine. However, from the context of the sentence, we should choose something that means the material besides the main topic is unnecessary. Choice *C*, irrelevant, means unimportant or unnecessary, so this is the best choice.

25. A: Choices *B*, *C,* and *D*, hyperbole, onomatopoeia, and simile, are all terms used in writing. Choice *B*, color, is specific to painting.

26. A: Chrissy is supposed to be a great team leader, so let's look at the words that are positive rather than negative. If someone *inspires* someone else then they are leading that person to encouragement and support, so *inspire* is the best answer. Confuse, daunt, and disable all denote negative actions, so these are not characteristics of a great team leader.

27. D: The students would act *sluggish*, or tired, when asked to perform one more test because of their exhaustion. The other words mean having energy, so these are incorrect.

28. C: The music was *rowdy*, which means loud or boisterous enough for someone to have to cover their ears. The other words denote calm or quiet, so they are incorrect.

29. A: *Bolted* means to run away quickly, so this is the best answer for someone who is stealing a purse.

30. B: *Mediocre* means average or inferior, so this is the best answer choice. The other words mean superiority, so they are incorrect, since an immature writing style means that the paper would not be superior.

31. B: The word *dismay* means being disappointed or showing stress, so this is the best word to use. Students who are supposed to give an unprepared speech will probably feel nervous about the situation, so the other answer choices, which all mean confident or happy, are incorrect.

32. D: Let's look for words that are most closely related to the word *extravagant*. Choice *B*, fancy, means extravagant and ornamental, so these are the best answers. Choices *A*, *C*, and *D* are versions of the words poor or cheap, so these are incorrect.

33. C: *Compensation* is the best word to put here because it means to repay or to make up for. If Rachel is sitting through an orientation for her job, she probably expects to be paid for that time she is learning.

34. A: The word *clarity* fits best here. Let's look at the context of the sentence. The second part of the sentence tells us that the group could see all the way down to the bottom of the river. This means that the water in the river must have been *clear*. *Clarity* means clearness, so this is our best answer choice.

Quantitative Reasoning

1. D: $\frac{59}{7}$

The original number was $8\frac{3}{7}$. Multiply the denominator by the whole number portion. Add the numerator and put the total over the original denominator.

$$\frac{(8 \times 7) + 3}{7} = \frac{59}{7}$$

2: D: $\frac{8}{7}$ is the same as $8 \times \frac{1}{7}$, which is represented by the first option: $\frac{1}{7} + \frac{1}{7} + \frac{1}{7} + \frac{1}{7} + \frac{1}{7} + \frac{1}{7} + \frac{1}{7} + \frac{1}{7}$. This can be thought of as cutting a pie into seven slices and then serving 8 slices. Since you need more slices than you have in your pie, you actually need to cut up two pies and take one piece from the second pie. This is because $\frac{8}{7}$ is an improper fraction, which means the numerator (top number) is greater than the denominator (bottom number).

3. B: The first step is to determine the unknown, which is in terms of the length, l.

The second step is to translate the problem into the equation using the perimeter of a rectangle:

$$P = 2l + 2w$$

The width is the length minus 2 centimeters. The resulting equation is:

$$2l + 2(l - 2) = 44$$

The equation can be solved as follows:

$2l + 2l - 4 = 44$	Apply the distributive property on the left side of the equation
$4l - 4 = 44$	Combine like terms on the left side of the equation
$4l = 48$	Add 4 to both sides of the equation
$l = 12$	Divide both sides of the equation by 4

The length of the rectangle is 12 centimeters. The width is the length minus 2 centimeters, which is 10 centimeters. Checking the answers for length and width forms the following equation:

$$44 = 2(12) + 2(10)$$

The equation can be solved using the order of operations to form a true statement: $44 = 44$.

4. A: In order to compare the fractions $\frac{4}{7}$ and $\frac{5}{9}$, a common denominator must be used. The least common denominator is 63, which is found by multiplying the two denominators together (7 × 9). The conversions are as follows:

$$\frac{4}{7} \times \frac{9}{9} = \frac{36}{63}$$

$$\frac{5}{9} \times \frac{7}{7} = \frac{35}{63}$$

Although they walk nearly the same distance, $\frac{4}{7}$ is slightly more than $\frac{5}{9}$ because $\frac{36}{63} > \frac{35}{63}$. Remember, the sign > means "is greater than." Therefore, Chris walks further than Tina, and Choice *A* correctly shows this expression in mathematical terms.

5. D: $3\frac{3}{5}$. Divide 54 by 15:

$$\begin{array}{r} 3 \\ 15\overline{)54} \\ -45 \\ \hline 9 \end{array}$$

The result is $3\frac{9}{15}$. Reduce the remainder for the final answer, $3\frac{3}{5}$.

6. B: Kareem arrived to his appointment at 8:45 a.m., since that's 15 minutes before 9:00 a.m., which was his scheduled time. He was taken back at 9:15 a.m., since 30 minutes after 8:45 a.m. is 9:15 a.m. His cleaning was 45 minutes, so he was done at 10:00 a.m.

7. C: The first step is to depict each number using decimals. $\frac{91}{100} = 0.91$

Dividing the numerator by denominator of $\frac{4}{5}$ to convert it into a decimal yields 0.80, while $\frac{2}{3}$ becomes 0.66 recurring. Rearrange each expression in ascending order, as found in Choice *C*.

8. D: First, calculate the difference between the larger value and the smaller value.

378 – 252 = 126

To calculate this difference as a percentage of the original value, and thus calculate the percentage *increase*, divide 126 by 252, then multiply by 100 to reach the percentage = 50%, Choice *D*.

9. B: To calculate the range in a set of data, subtract the highest value with the lowest value. In this graph, the range of Mr. Lennon's students is 5, which can be seen physically in the graph as having the smallest difference compared with the other teachers between the highest value and the lowest value.

10. B: To calculate the difference between the two scores for Student 3, subtract the score from Mr. Taylor's class (77) from the score in Mr. O'Shea's class (85): $85 - 77 = 8$.

11. C: The number of days can be found by taking the total amount Bernard needs to make and dividing it by the amount he earns per day:

$$\frac{300}{80} = \frac{30}{8} = \frac{15}{4} = 3.75$$

But Bernard is only working full days, so he will need to work 4 days, since 3 days is not a sufficient amount of time.

12. D: Using the key, it can be seen that each fruit symbol is equivalent to 2 counts of that fruit. There are 7 bananas pictured, which means 14 bananas were eaten because 7 × 2 = 14. There were 6 pineapples eaten because 3 are pictured and 3 × 2 = 6. Then, the difference must be found: 14 – 6 = 8, so 8 more bananas were eaten. Therefore, Choice *D* is the correct answer.

13. B: The place value to the right of the hundredth place, which would be the thousandth place, is what gets used. The value in the hundredth place is 7. The number in the place value to its right is greater than 4, so the 7 gets bumped up to 8. Everything to its right turns to a zero, to get 423.2800. The zeros are dropped because it is part of the decimal.

14. C: These labels correctly describe a real-world application of the input-output table shown. The number of tricycles would need to be multiplied by 3 (the number of wheels on a tricycle) in order to find the number of total wheels in a store's inventory. Choice *A* is not a correct modeling of a real-world situation. A stable chair would have 4 legs, not 3. Choice *B* is incorrect as it mixes up the number of wheels on a tricycle with the number of tricycles. The number of wheels cannot be the variable (changing) item for this calculation. Choice *D* does something similar as Choice *B*, by mixing up the variable and the multiplier; dogs would have a set number of paws, not one that would change.

15. A: Lining up the given scores provides the following list: 60, 75, 80, 85, and one unknown. Because the median needs to be 80, it means 80 must be the middle data point out of these five. Therefore, the unknown data point must be the fourth or fifth data point, meaning it must be greater than or equal to 80. The only answer that fails to meet this condition is 60.

16. A: Let the unknown score be *x*. The average will be:

$$\frac{5 \cdot 50 + 4 \cdot 70 + x}{10} = \frac{530 + x}{10} = 55$$

Multiply both sides by 10 to get $530 + x = 550$, or $x = 20$.

17. C: For manufacturing costs, the base cost, which is added, is \$50,000, while the cost per unit is \$40 times the number of saws. So, the answer is $40x + 50{,}000$.

18. C: A die has an equal chance for each outcome. Since it has six sides, each outcome has a probability of $\frac{1}{6}$. The chance of a 1 or a 2 is therefore $\frac{1}{6} + \frac{1}{6} = \frac{1}{3}$.

19. C: 85% of a number means multiplying that number by 0.85. So, $0.85 \times 20 = \frac{85}{100} \times \frac{20}{1}$, which can be simplified to $\frac{17}{20} \times \frac{20}{1} = 17$.

20. B: The dot plot in Choice *B* is correct because, like the table, it shows that 7 students play soccer, 1 swims, 3 run track, 6 play basketball, 6 play baseball, and 2 play tennis. Each dot represents one student, just like one hash mark in the table represents one student.

21. C: We are trying to find x, the number of red cans. The equation can be set up like this:

$$x + 2(10 - x) = 16$$

The left x is actually multiplied by \$1, the price per red can. Since we know Jessica bought 10 total cans, $10 - x$ is the number blue cans that she bought. We multiply the number of blue cans by \$2, the price per blue can.

That should all equal \$16, the total amount of money that Jessica spent. Working that out gives us:

$$x + 20 - 2x = 16$$

$$20 - x = 16$$

$$x = 4$$

22. C: Each hour on the clock represents 30 degrees. For example, 3:00 represents a right angle. Therefore, 5:00 represents 150 degrees.

23. B: The number negative four is classified as a real number because it exists and is not imaginary. It is rational because it does not have a decimal that never ends. It is an integer because it does not have a fractional component. The next classification would be whole numbers, for which negative four does not qualify because it is negative. Although -4 could technically be considered a complex number because complex numbers can have either the real or imaginary part equal zero and still be considered a complex number, Choice *A* is wrong because -4 is not considered an irrational number because it does not have a never-ending decimal component.

24. B: Every 8 mL of medicine requires 5 mL. The 45 mL first needs to be split into portions of 8 mL. This results in $\frac{45}{8}$ portions. Each portion requires 5 mL. Therefore, $\frac{45}{8} \times 5 = \frac{45*5}{8} = \frac{225}{8}$ mL is necessary.

25. B: According to the order of operations, multiplication and division must be completed first from left to right. Then, addition and subtraction are completed from left to right. Therefore:

$$7 \times 7 \div 7 + 7 - 7 \div 7$$

$$49 \div 7 + 7 - 7 \div 7$$

$$7 + 7 - 7 \div 7$$

$$7 + 7 - 1$$

$$14 - 1 = 13$$

26. B: This inequality can be seen with the use of a number line. $\frac{3}{7}$ is close to $\frac{1}{2}$. $\frac{5}{6}$ is close to 1, but less than 1, and $\frac{8}{7}$ is greater than 1. Therefore, $\frac{3}{7}$ is less than $\frac{5}{6}$.

27. C: The triangle in Choice *B* doesn't contain a line of symmetry. The figures in Choices *A* and *D* do contain a line of symmetry but it is not the line that is shown here. Choice *C* is the only one with a correct line of symmetry shown, such that the figure is mirrored on each side of the line.

28. B: The sample space is made up of $8 + 7 + 6 + 5 = 26$ balls. The probability of pulling each individual ball is $\frac{1}{26}$. Since there are 7 yellow balls, the probability of pulling a yellow ball is $\frac{7}{26}$.

29. D: The equation is $3(n + 7) \geq 32$. 3 times the sum of a number and 7 is greater than or equal to 32 can be translated into equation form utilizing mathematical operators and numbers.

30. D: $350,000: Since the total income is $500,000, then a percentage of that can be found by multiplying the percent of Audit Services as a decimal, or 0.40, by the total of 500,000. This answer is found from the equation:

$$500000 \times 0.4 = 200000$$

The total income from Audit Services is $200,000.

For the income received from Taxation Services, the following equation can be used:

$$500000 \times 0.3 = 150000$$

The total income from Audit Services and Taxation Services is:

$$150{,}000 + 200{,}000 = 350{,}000$$

Another way of approaching the problem is to calculate the easy percentage of 10%, then multiply it by 7, because the total percentage for Audit and Taxation Services was 70%. 10% of 500,000 is 50,000. Then multiplying this number by 7 yields the same income of $350,000.

31. B: 182 is in the millions, 36 is in the thousands, 421 is in the hundreds, and 356 is the decimal.

32. D: 128

This question involves the percent formula.

$$\frac{32}{x} = \frac{25}{100}$$

We multiply the diagonal numbers, 32 and 100, to get 3,200. Dividing by the remaining number, 25, gives us 128.

The percent formula does not have to be used for a question like this. Since 25% is ¼ of 100, you know that 32 needs to be multiplied by 4, which yields 128.

33. D: A mixed number contains both a whole number and either a fraction or a decimal. Therefore, the mixed number is $16\frac{1}{2}$.

34. D: The slope from this equation is 50, and it is interpreted as the cost per gigabyte used. Since the *g*-value represents number of gigabytes and the equation is set equal to the cost in dollars, the slope relates these two values. For every gigabyte used on the phone, the bill goes up 50 dollars.

35. C: When giving an answer to a math problem that is in fraction form, it always should be simplified. Both 3 and 15 have a common factor of 3 that can be divided out, so the correct answer is:

$$\frac{3 \div 3}{15 \div 3} = \frac{1}{5}$$

36. B: The number line shows:

$$x > -\frac{3}{4}$$

Each inequality must be solved for x to determine if it matches the number line. Choice *A* of $4x + 5 < 8$ results in $x < -\frac{3}{4}$, which is incorrect. Choice *C* of $-4x + 5 > 8$ yields $x < -\frac{3}{4}$, which is also incorrect. Choice *D* of $4x - 5 > 8$ results in $x > \frac{13}{4}$, which is not correct. Choice B, $-4x + 5 < 8$ is the only choice that results in the correct answer of:

$$x > -\frac{3}{4}$$

37. D: The two lines are neither parallel nor perpendicular. Parallel lines will never intersect or meet. Therefore, the lines are not parallel. Perpendicular lines intersect to form a right angle (90°). Although the lines intersect, they do not form a right angle, which is usually indicated with a box at the intersection point. Therefore, the lines are not perpendicular.

38. A: If each man gains 10 pounds, every original data point will increase by 10 pounds. Therefore, the man with the original median will still have the median value, but that value will increase by 10. The smallest value and largest value will also increase by 10 and, therefore, the difference between the two won't change. The range does not change in value and, thus, remains the same.

Reading Comprehension

1. D: Caribbean island destinations. The topic of the passage can be described in a one- or two-word phrase. Choices *A*, *B*, and *C* are all mentioned in the passage. However, they are too vague to be considered the main topic of the passage.

2. C: Family resorts and activities. Remember that supporting details help readers find out the main idea by answering questions like *who*, *what*, *where*, *when*, *why*, and *how*. In this question, beaches, cruising to the Caribbean, and local events are not talked about in the passage. However, family resorts and activities are talked about.

3. D: Opinion. An opinion is when the author states their own thoughts on a subject. In this sentence, the author says that the reader will not regret the vacation. The author says that it may be the best and most relaxing vacation. But this may not be true for the reader. Therefore, the statement is the author's opinion. Facts would have evidence, like that collected in a science experiment.

4. B: Persuade readers. The author is trying to persuade readers to go to a Caribbean island destination by giving the reader fun facts and a lot of fun options. Not only does the author give a lot of details to support their opinion, the author also implies that the reader would be "wrong" if they didn't want to visit a Caribbean island. This means the author is trying to persuade the reader to visit a Caribbean island.

5. D: Swimsuit or swim trunks and goggles. The passage mentions visitors can swim in crystal blue waters and swim with dolphins, so anything having to do with swimming is necessary for a Caribbean island vacation. The other choices are incorrect because the passage does not mention sports, camping, or cold temperatures.

6. A: Rain. Rain is the cause in this passage because it is why something happened. The effects are plants growing, rainbows, and puddle jumping.

7. A: Comparing. The author is comparing the plants, trees, and flowers. The author is showing how these things react the same to rain. They all get important nutrients from rain. If the author described the differences, then it would be contrasting, Choice *B*.

8. C: Necessary. *Vital* can mean different things depending on the context or how it is used. But in this sentence, the word *vital* means necessary. The word *vital* means full of life and energy. Choice *A*, *dangerous*, is almost an antonym for the word we are looking for since the sentence says the nutrients are needed for growing. Something needed would not be dangerous. Choices *B* and *D*, *energetic* and *truthful*, do not make sense. The best context clue is that it says the vital nutrients are needed, which tells us they are necessary.

9. B: Rainbows. This passage mentions several effects. Effects are the outcome of a certain cause. Remember that the cause here is rain, so Choice *A* is incorrect. Choice *B* makes the most sense because the effects of the rain in the passage are plants growing, rainbows, and puddle jumping. Choice *C*, weather, is not an effect of rain but describes rain in a general sense. Since the cause is rain, Choice *D*—brief spurts of sunshine—doesn't make sense because rain doesn't *cause* brief spurts of sunshine.

10. D: Puddle jumping can be done in or after the rain. The passage mentions that puddle jumping is a fun activity that is available while it's raining, therefore Choice *D* is correct. Choice *A* is incorrect, because even though the passage mentions that rain can put a damper on the day, it is not a reason that

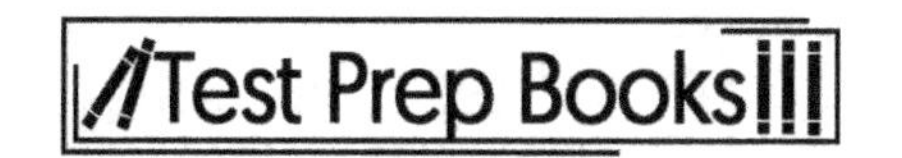

it is helpful or fun. Choices *B* and *C* are incorrect because the opposite is true. Rain helps plants grow and reflects and refracts light, creating rainbows.

11. D: Siberian huskies are great pets but require a lot of time and energy. The author mentions that Siberian huskies are so friendly, that they would most likely greet a robber that came into their house. This certainly does not make them good guard dogs. In the passage, the writer explains that huskies dig a lot, they need lots of brushing, and they need to be taken on long walks, which makes them not always easy to care for. Therefore, Siberian huskies are great pets but require a lot of time and energy.

12. A: Biased: The author may be biased because they show that they like one dog breed over another in an unfair way. Hasty, Choice *B*, means quick to judge. Choice *C*, impartial, is the opposite of biased and means very fair, without being opinionated. Irrational, Choice *D*, means something that doesn't make sense.

13. D: It is an experienced dog owner. We do not have any clues from the paragraph whether the narrator is a man or a woman, so Choices *A* and *B* are incorrect. Also, the narrator talks about having other dogs before, so they cannot be a new dog owner. Choice *D* makes sense because the narrator talks about having other dogs before Lola, which means that they have been a dog owner before.

14. B: Amiable. Amiable means friendly, and the passage mentions that Lola loves making friends. Choice *A,* aggressive, means hostile or threatening, which is incorrect because the author explains a robber would most likely be greeted with kisses and a tail wag. Choice *C*, anxious, and Choice *D,* mean worried or uneasy and are not supported in the paragraph.

15. A: Choice *A* is a photo of a Siberian Husky like Lola. Test takers do not need to be familiar with different dog breeds to correctly answer this question. Instead, they can be detectives and use clues from the passage about what Lola looks like. For one, the narrator mentions Lola's long fur, which sounds bushy and full because it has to be brushed so much! Dogs in Choices *B* and *C* (the Chihuahua and the Dalmatian) are ruled out because of their short hair. The narrator also mentions that Lola has a "wolf-like appearance" that may scare a robber. Even though Choice *D* (a Shih-Tzu) has very long hair, that dog does not look like a wolf. Furthermore, Choices *B* and *C* do not look like wolves.

16. D: Readers should focus on the details in the passage to answer this question. The beginning of the passage, as well as the main idea, states that writing ten-minute plays may *seem* difficult, but it actually isn't. Therefore, Choice *A* is incorrect. Choices *B* and *D* are opposites. The passage mentions how descriptive details *are* important to help set the mood, tone, and theme, so Choice *B* is incorrect, and Choice *D* is the best answer. Lastly, it is said that the theme is set in the descriptive details. The theme should come right at the beginning of the play and not the climax, so Choice *C* is incorrect.

17. A: Pictures can tell stories as well as, if not better than, words. This is a phrase used a lot in the English language. In the case of a ten-minute play, playwrights would be smart to use images to cut down on the dialogue, since ten minutes is not a long time. This passage was all about how writing a short play isn't actually that hard even for a new playwright. The author of the passage persuades readers by stating that pictures make it a lot simpler.

18. D: The protagonist of a story is the main character. Without knowing this, test takers can try to find the correct choice by using clues from the passage. The passage states: *As the play progresses, the protagonist must cross the point of no return in some way; this is the climax of the story. Then, as in a written story, you create a resolution to the life-changing event of the protagonist.* This information should help rule out Choices *B* and *C* (student and playwright) since it is clear that the protagonist is in the play. Choice *A*, actor, may be an attractive choice since an actor is in the play, but careful readers will notice that is says *the protagonist* meaning there is only one. There are likely multiple actors in the play, because the passage mentions dialogue, which must include at least two people.

19. B: Choice *B* is correct because of the opening statement: "Learning how to write a ten-minute play may seem like a monumental task at first; but, if you follow a simple creative writing strategy, similar to writing a narrative story, you will be able to write a successful drama." The passage does not talk about how long a playwright spends doing revisions and rewrites. So, Choice *C* is incorrect. None of the other choices are supported by points in the passage.

20. B: According to the passage, the resolution happens after the climax and is the ending to the play. Therefore, Choice *B* is correct. Choice *A,* cause, comes at the beginning of the play, like a cause and effect. Choice *C*, conversation, is another word for dialogue. While Choice *D*, courage, is a synonym for resolution and bravery, that is not how it is being used in this passage.

21. D: *Astonishing* most closely means surprising. Readers may be tempted to choose *celebrated,* Choice *A*, because the passage mentions a lot of celebrating. However, using clues from the sentence that the other team always wins would help make *surprising* a better choice. Choice *B*, "a close win," does not make sense because the narrator does not say anything about the actual score. Choice *C*, expected, can be ruled out because the same sentence mentions that the other team wins every year.

22. D: The passage talks about Little League and Max scoring the winning run. These are clues that it is about baseball. So, Choice *A* is incorrect. The main idea of the story is about the baseball team winning the championships. It is true that the team eats at the diner and one player is a vegetarian. However, these are supporting details.

23. A: This question was a little tricky because both *B* and *C* seem like they could be true. Nervous and disappointed, Choices *A* and *D*, should be easy to rule out because the narrator was happy about the win. The passage does mention that the other team usually wins, and it has been twenty years since the narrator's team won. However, these are just details. Choice *D*, informational, is not really a tone or mood. It refers to a type of writing that is educational. This passage is a story with excited emotion.

24. B: Readers are told that Max is a vegetarian. This means he does not eat meat. Bacon is a meat, so we can guess that Max does not eat bacon. Readers may be tempted to choose Choice *A*, that Max is overweight, because it mentions that he follows a specific diet (vegetarian). However, he is active and in sports and got the winning run. Therefore, Choice *B* makes more sense than Choice *A*. Choice *C* is incorrect because the narrator mentions Max by name and doesn't say "I," so the narrator is not Max. We can also infer that Max has not been on the team for twenty years, because he would be too old!

25. B: Dedication. Even though the team was down the whole game until the final inning, the team was dedicated to winning the championship game, so Choice *B* is correct. Choice *A*, decision, is incorrect, because the team did not make a choice in this passage. Choices *C* and *D*, laziness and weakness, are antonyms of perseverance.

Math Achievement

1. D: The situation can be described by the equation $? \times 2$. Filling in for the missing numbers would result in $3 \times 2 = 6$ and $7 \times 2 = 14$. Therefore, the missing numbers are 6 and 14. The other choices are miscalculations or misidentification of the pattern formed by the table.

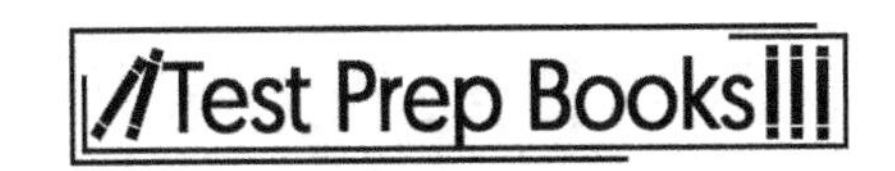

2. D: Your total expenses were \$77 and your total earnings were \$95, so you earned more money than you spent. Therefore, you were able to balance your budget successfully.

3. B: First, subtract 4 from each side. This yields: $6t = 12$. Now, divide both sides by 6 to obtain $t = 2$.

4. D: $27\frac{7}{22}$

Set up the division problem.

$$44\overline{)1202}$$

44 does not go into 1 or 12 but will go into 120 so start there.

$$\begin{array}{r} 27 \\ 44\overline{)1202} \\ -88 \\ \hline 322 \\ -308 \\ \hline 14 \end{array}$$

The answer is $27\frac{14}{44}$.

Reduce the fraction for the final answer.

$$27\frac{7}{22}$$

5. D: 100,000.0 cm. To convert from kilometers to centimeters, move the decimal 5 places to the right.

$$1.0\ km = 100000.0\ cm$$

6. B: When you add 78.654 and 4.900 you get 83.554. Because the number in the thousandths place is not greater than four, the number in the hundredths would just stay the same. The total is: 83.55.

7. B: 3.6

Divide 3 by 5 to get 0.6 and add that to the whole number 3, to get 3.6. An alternative is to incorporate the whole number 3 earlier on by creating an improper fraction: $\frac{18}{5}$. Then dividing 18 by 5 to get 3.6.

8. C: 104,165

Set up the problem and add each column, starting on the far right (ones). Add, carrying anything over 9 into the next column to the left. Solve from right to left.

9. C:

$9x + x - 7 = 16 + 2x$	Combine $9x$ and x.
$10x - 7 = 16 + 2x$	
$10x - 7 + 7 = 16 + 2x + 7$	Add 7 to both sides to remove (-7).
$10x = 23 + 2x$	
$10x - 2x = 23 + 2x - 2x$	Subtract 2x from both sides to move it to the other side of the equation.
$8x = 23$	
$\frac{8x}{8} = \frac{23}{8}$	Divide by 8 to get x by itself.
$x = \frac{23}{8}$	

10. B: 12

Calculate how many gallons the bucket holds.

$$11.4\ L \times \frac{1\ gal}{3.8\ L} = 3\ gal$$

Now how many buckets to fill the pool which needs 35 gallons.

$$35/3 = 11.67$$

Since the amount is more than 11 but less than 12, we must fill the bucket 12 times.

11. B: The team needs a total of $270, and each box earns them $3. Therefore, the total number of boxes needed to be sold is $270 \div 3$, which is 90. With 15 people on the team, the total of 90 can be divided by 15, which equals 6. This means that each member of the team needs to sell 6 boxes for the team to raise enough money to buy new uniforms.

12. B: 648.77

Set up the problem, with the larger number on top and numbers lined up at the decimal. Insert 0 in any blank spots to the right of the decimal as placeholders. Begin subtracting with the far-right column. Borrow 10 from the column to the left, when necessary.

13. C: The range of the entire stem-and-leaf plot is found by subtracting the lowest value from the highest value, as follows: $59 - 20 =$ 39 cm. All other choices are miscalculations read from the chart.

14. C: To calculate the total greater than 35, the number of measurements above 35 must be totaled; 36, 37, 37, 47, 49, 54, 56, 59 = 8 measurements. Choice *A* is the number of measurements in the 3 categories, Choice *B* is the number in the 4 and 5 categories, and Choice *D* is the number in the 3, 4, and 5 categories.

15. C: $\frac{4}{3}$

Set up the problem and find a common denominator for both fractions.

$$\frac{14}{33} + \frac{10}{11}$$

Multiply each fraction across by 1 to convert to a common denominator

$$\frac{14}{33} \times \frac{1}{1} + \frac{10}{11} \times \frac{3}{3}$$

Once over the same denominator, add across the top. The total is over the common denominator.

$$\frac{14 + 30}{33} = \frac{44}{33}$$

Reduce by dividing both numerator and denominator by 11.

$$\frac{44 \div 11}{33 \div 11} = \frac{4}{3}$$

16. D: This situation can be modeled using equivalent ratios. Since the ratio of boys to girls is 3 to 2, then the total for that ratio is 5. The ratio of boys to total kids is 3 to 5. Using these numbers, the following equation can be written:

- $\frac{3}{5} = \frac{x}{30}$

Since 30 is the given total number of kids in the class, the value of x represents the number of boys. To solve this equation, cross-multiplication can be used. This turns the equation into:

$$3 \times 30 = 5 \times x$$

Next, solve this equation in the following steps:

- $90 = 5x$
- $18 = x$

There are 18 boys in the kindergarten class.

To check that the ratios are correct, subtraction can be used to find the number of girls in the class: $30 - 18 = 12$. The ratio of boys to girls is 18 to 12, which can be reduced to 3 to 2.

17. D: This problem can be solved by setting up a proportion involving the given information and the unknown value. The proportion is:

$$\frac{21\ pages}{4\ nights} = \frac{140\ pages}{x\ nights}$$

Solving the proportion by cross-multiplying, the equation becomes $21x = 4 \times 140$, where $x = 26.67$. Since it is not an exact number of nights, the answer is rounded up to 27 nights. Twenty-six nights would not give Sarah enough time.

18. B: 110,833

Set up the problem, with the larger number on top. Begin subtracting with the far-right column (ones). Borrow 10 from the column to the left, when necessary.

19. C: The first step in solving this problem is finding the total number of students in the school. The sum of students is $550 + 635 = 1185$. Out of the 1185 total students, 635 of them are girls. To find the percentage of girls, 635 can be divided by 1185. This division yields a decimal of 0.5358. Multiplying this number by 100 turns it into a percentage of 53.6.

20: C. The two points are at -5 and 0 for the x-axis and at -3 and at -1 for y-axis respectively. Therefore, the two points have the coordinates of (-5, -3) and (0, -1).

21. B: The long division would be completed as follows:

```
     24
36|864
  - 72↓
   144
```

22. D: Your total expenses were $145 and your total earnings were $140, so you spent more than you earned. Therefore, you need to borrow money from your savings account to cover your total expenses. Hopefully, next month, you will be able to place that borrowed money back into your account.

23. B: Setting up a proportion is the easiest way to represent this situation. The proportion becomes $\frac{20}{x} = \frac{40}{100}$, where cross-multiplication can be used to solve for x. The answer can also be found by observing the two fractions as equivalent, knowing that twenty is half of forty, and fifty is half of one-hundred.

24. D: The average test grade can be found by adding up all 4 grades and dividing the sum by 4. Since 3 of the grades are given and 1 is missing, an equation can be used to solve for the unknown. The equation to find the average is $\frac{87+85+78+x}{4} = 85$. Multiplying both sides by 4 yields the equation $87 + 85 + 78 + x = 340$. Adding up the test grades that are known and subtracting the sum from 340 yields the missing test grade, which is 90.

25. B: The total number of coins in the jar is 86, which is the sum of all the coins. The probability of Nina choosing a coin other than a penny or a dime can be found by calculating the total of quarters and nickels. This total is 31. Taking 31 and dividing it by 86 gives the probability of choosing a coin that is not a penny or a dime. This decimal found from the fraction $\frac{31}{86}$ is 0.36.

26. A: For the first card drawn, the probability of a King being pulled is $\frac{4}{52}$. Since this card isn't replaced, if a King is drawn first the probability of a King being drawn second is $\frac{3}{51}$. The probability of a King being drawn in both the first and second draw is the product of the two probabilities: $\frac{4}{52} \times \frac{3}{51} = \frac{12}{2652}$. This fraction, when divided by 12, equals $\frac{1}{221}$.

27. D: The expression is three times the sum of twice a number and 1, which is $3(2x + 1)$. Then, 6 is subtracted from this expression.

28. C: The first step in solving this equation is to collect like terms on the left side of the equation. This yields the new equation $-4 + 8x = 8 - 10x$. The next step is to move the x-terms to one side by adding 10 to both sides, making the equation $-4 + 18x = 8$. Then the -4 can be moved to the right side of the equation to form $18x = 12$. Dividing both sides of the equation by 18 gives a value of 0.67, or $\frac{2}{3}$.

29. D: Denote the width as *w* and the length as *l*. Then, $l = 3w + 5$. The perimeter is $2w + 2l = 90$. Substituting the first expression for *l* into the second equation yields $2(3w + 5) + 2w = 90$, or $8w = 80$, so$wl = 10$. Putting this into the first equation, it yields:

$$l = 3(10) + 5 = 35$$

30. D: Because there are about 5km in 3 miles, if he runs 15 miles, which is 3 miles x 5, he runs 5 km x 5, or 25 km.

Practice Test #2

Verbal Reasoning

Synonyms

Each of the questions below has one word. The one word is followed by four words or phrases. Please select one answer whose meaning is closest to the word in capital letters.

1. INITIATE:
   a. begin
   b. sedate
   c. surmise
   d. vacillate

2. PASSIVE:
   a. compatible
   b. indifferent
   c. snub
   d. tranquil

3. ROBUST:
   a. abort
   b. sturdy
   c. verbose
   d. weak

4. PRAGMATIC:
   a. fickle
   b. indignant
   c. sensible
   d. strenuous

5. ALOOF:
   a. detached
   b. envious
   c. jubilant
   d. limber

6. JUVENILE:
   a. fatuous
   b. immature
   c. impoverished
   d. stable

7. COURTEOUS:
    a. extravagant
    b. facile
    c. meager
    d. polite

8. MATRIMONY:
    a. entity
    b. hiatus
    c. marriage
    d. parody

9. PACIFY:
    a. patent
    b. soothe
    c. stagnate
    d. trek

10. PRUDENT:
    a. careful
    b. impatient
    c. likeable
    d. tranquil

11. COLLEAGUE:
    a. college
    b. coworker
    c. executive
    d. subordinate

12. ESSENTIAL:
    a. belligerent
    b. mundane
    c. necessary
    d. pervasive

13. ORATOR:
    a. dancer
    b. singer
    c. speaker
    d. writer

14. SCORN:
    a. cheer
    b. esteem
    c. praise
    d. ridicule

15. CHERISH:
    a. admonish
    b. adore
    c. bombard
    d. command

16. ASTOUND:
    a. amaze
    b. espy
    c. promote
    d. regard

17. LUCRATIVE:
    a. apathetic
    b. productive
    c. singular
    d. supreme

Sentence Completion

Select the word or phrase that most correctly completes the sentence.

18. Space is often referred to as the great __________.
    a. around
    b. diffuse
    c. planet
    d. void

19. The thievery merited __________ punishment.
    a. cut
    b. foul
    c. severe
    d. shake

20. The toddler, who had just learned to speak, seemed rather __________.
    a. cranky
    b. quiet
    c. silent
    d. verbose

21. The bullies __________ the younger boy, causing him to feel worthless.
    a. disparaged
    b. fixed
    c. praised
    d. uplifted

22. Though the valedictorian was very smart, he was too __________ to have very many friends.
    a. agrarian
    b. altruistic
    c. amicable
    d. egotistic

23. The car was jostled by the rocky __________.
    a. arboreal
    b. celestial
    c. firmament
    d. terrain

24. The king's army was able to easily __________ his brother's army, and quickly crushed the rebellion.
    a. avoid
    b. brazen
    c. encompass
    d. retreat

25. After sleeping in his car for the past few months, the tiny hotel room seemed like a mansion by __________.
    a. contrast
    b. improve
    c. partisan
    d. revive

26. The babysitter was constantly stressed out when watching Nathaniel. The young boy seemed to be capable of nothing, other than his uncanny ability to __________ his babysitter.
    a. breathe
    b. exasperate
    c. humor
    d. stifle

27. The rebels fought in order to __________ their brothers from the evil dictator.
    a. agitate
    b. fracture
    c. instigate
    d. liberate

28. The little girl stared __________ after her balloon as it floated away.
    a. dejectedly
    b. jubilantly
    c. passively
    d. petrified

29. The starving man was disheartened when he reached the summit of the hill and realized that only a __________ wasteland awaited him.
    a. barren
    b. fruitful
    c. lavish
    d. sumptuous

30. When she saw the crayon drawings on the wall, the mother had no choice but to __________ her sons.
   a. chastise
   b. choose
   c. honor
   d. locate

31. When the boy saw how sincere the girl's apology was, he decided to __________ her of her faults.
   a. acquire
   b. forgive
   c. quit
   d. stall

32. The mountain's __________ made it difficult to breathe in the very high atmosphere.
   a. altitude
   b. behavior
   c. haughtiness
   d. outlook

33. Annie, fuming inside, was __________ when her turn in line was passed up.
   a. amiable
   b. innocent
   c. irate
   d. tangible

34. The children did not mean to __________ the bear with their loud talking and jumping up and down, but they were very excited.
   a. damage
   b. inspire
   c. provoke
   d. soothe

Quantitative Reasoning

1. If you were showing your friend how to round 245.867 to the nearest tenth, which place value would be used to decide whether to round up or round down?
   a. Hundredths
   b. Tenths
   c. Ones
   d. Hundreds

2. A teacher is showing students how to evaluate $5 \times 6 + 4 \div 2 - 1$. Which operation should be completed first?
   a. Addition
   b. Division
   c. Multiplication
   d. Subtraction

3. What is the definition of a factor of the number 36?
    a. A prime number that is multiplied times 36
    b. An even number that is multiplied times 36
    c. A number that can be divided by 36 and have no remainder
    d. A number that 36 can be divided by and have no remainder

4. Which represents the number 0.65 on a number line?

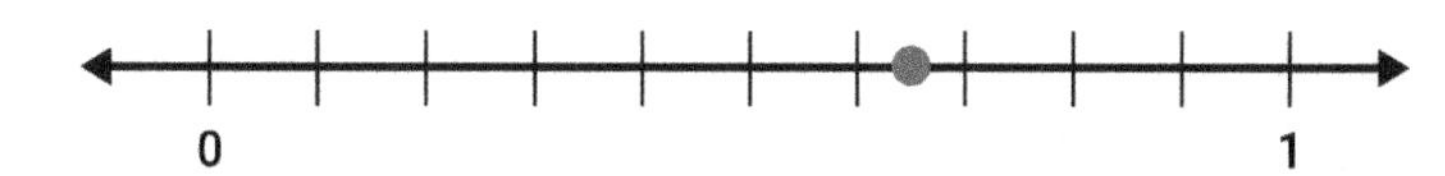

a.

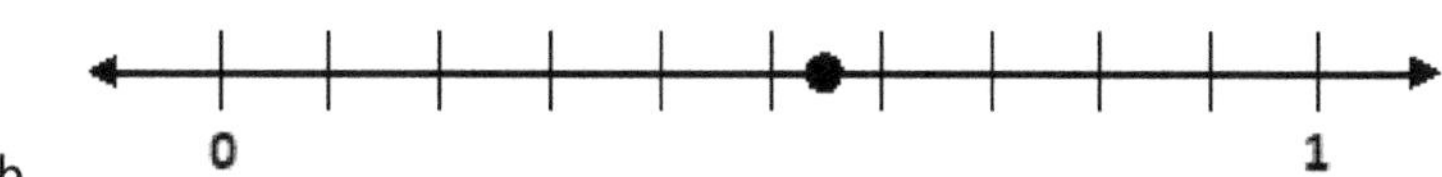

b.

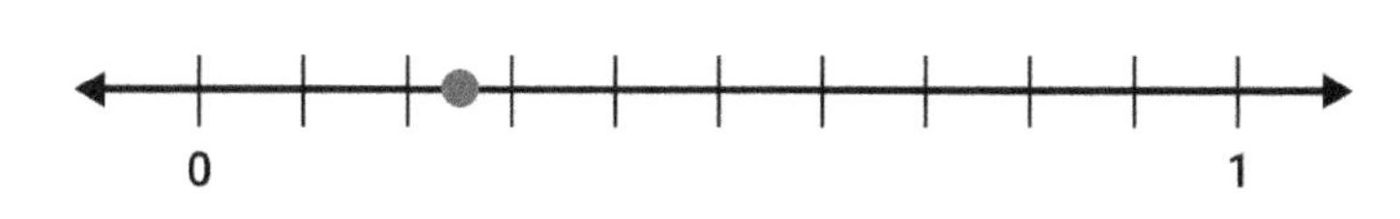

c.

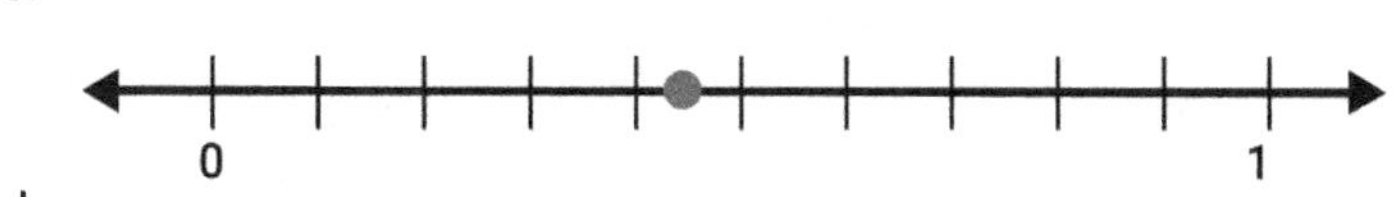

d.

5. Which of the following numbers is greater than (>) 220,058?
    a. 220,158
    b. 202,058
    c. 220,008
    d. 217,058

6. Which of the following is not a parallelogram?

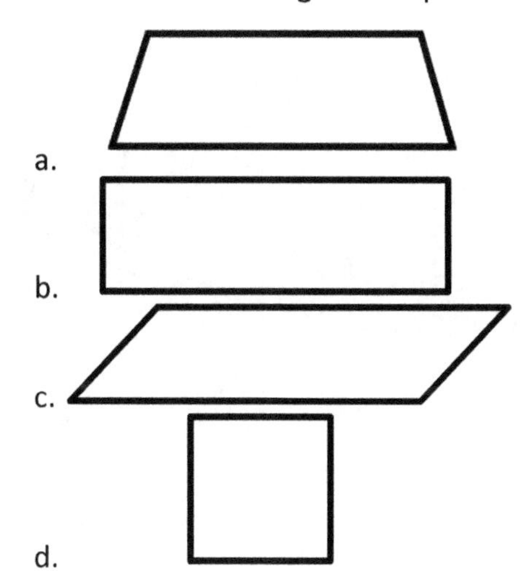

a.
b.
c.
d.

7. What equation, involving the addition of two fractions, is represented on the following number line?

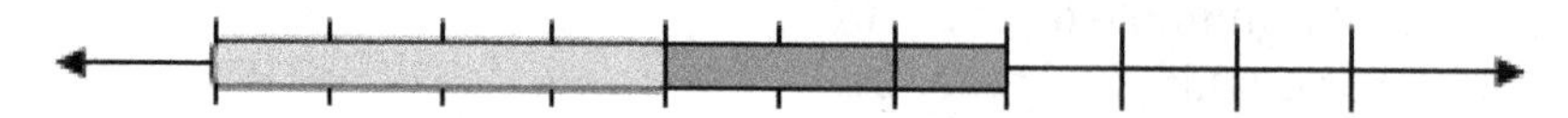

a. $\frac{4}{5} + \frac{3}{5} = \frac{7}{5}$
b. $\frac{4}{5} + \frac{7}{5} = \frac{7}{5}$
c. $\frac{3}{5} + \frac{3}{5} = \frac{6}{5}$
d. $\frac{4}{5} + 1\frac{3}{5} = \frac{7}{5}$

8. What is the product of 26×12?
a. 78
b. 202
c. 302
d. 312

9. What two fractions add up to $\frac{7}{6}$?
a. $\frac{2}{3} + \frac{5}{3}$
b. $\frac{1}{5} + \frac{6}{5}$
c. $\frac{1}{6} + \frac{6}{6}$
d. $\frac{1}{2} + \frac{6}{4}$

10. What is the measure of Angle *B* to the nearest degree displayed on the following protractor?

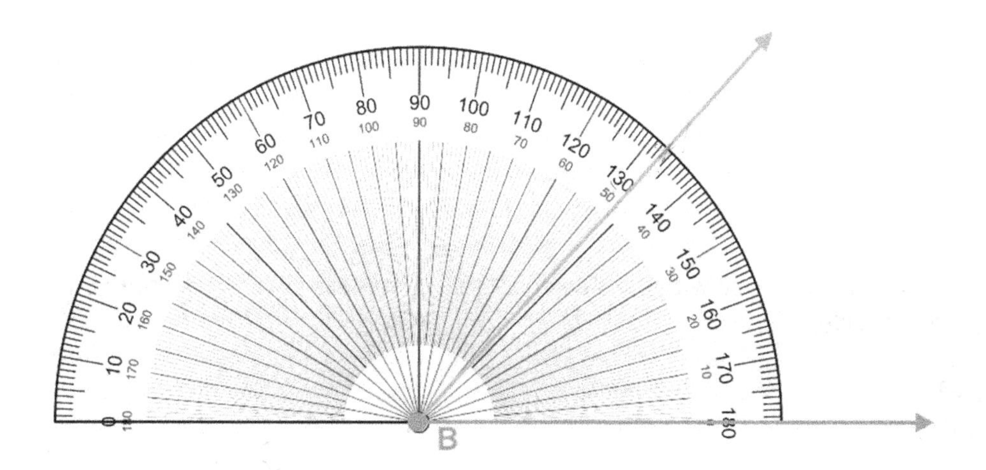

a. 47°
b. 53°
c. 133°
d. 136°

11. Students should line up decimal places within the given numbers before performing which of the following?
a. Division
b. Exponents
c. Multiplication
d. Subtraction

12. In an office, there are 50 workers. A total of 60% of the workers are women, and the chances of a woman wearing a skirt is 50%. If no men wear skirts, how many workers are wearing skirts?
a. 12
b. 15
c. 16
d. 20

13. Subtract and express in reduced form $\frac{43}{45} - \frac{11}{15}$.
   a. $\frac{2}{9}$
   b. $\frac{10}{45}$
   c. $\frac{16}{15}$
   d. $\frac{32}{30}$

14. What is the volume of a cube with the side equal to 5 centimeters? ($V = s^3$ where V is volume and s is the length of one side)
   a. 10 cm^3
   b. 15 cm^3
   c. 50 cm^3
   d. 125 cm^3

15. What is the value, to the nearest tenths place, of the point indicated on the following number line?

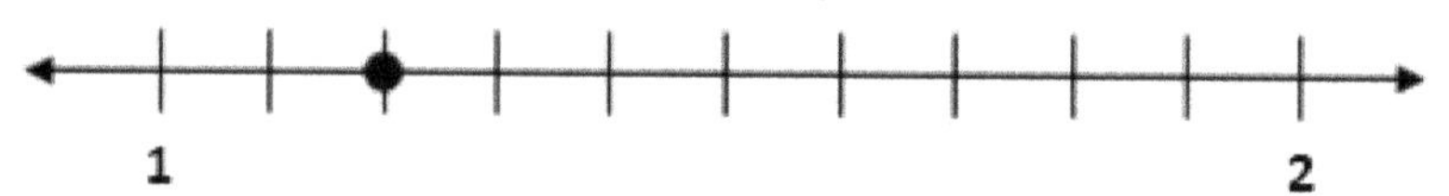

   a. 0.2
   b. 1.2
   c. 1.4
   d. 2.2

16. A rectangle was formed out of pipe cleaner. Its length was $\frac{1}{2}$ feet and its width was $\frac{11}{2}$ inches. What is its area in square inches?
   a. $\frac{11}{4}$ $inch^2$
   b. $\frac{11}{2}$ $inch^2$
   c. 22 $inch^2$
   d. 33 $inch^2$

17. A teacher cuts a pie into 6 equal pieces and takes one away. What topic would she be introducing to the class by using such a visual?
   a. Addition
   b. Decimals
   c. Fractions
   d. Measurement

18. Which of the following equations is correct?
   a. $123 \div 4 = 33$
   b. $123 \div 4 = 30\ R3$
   c. $123 \div 4 = 3\ R30$
   d. $123 \div 4 = 30 + 3$

19. This chart indicates how many sales of CDs, vinyl records, and MP3 downloads occurred over the last year. Approximately what percentage of the total sales was from CDs?

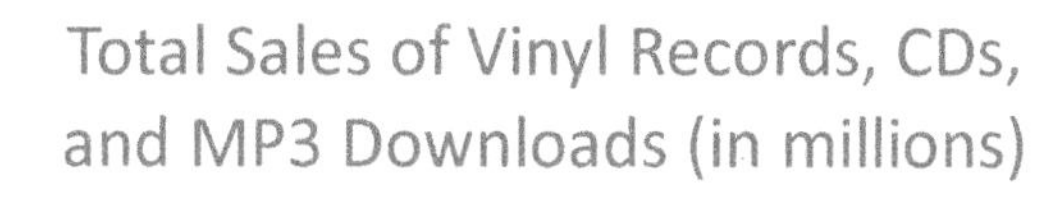

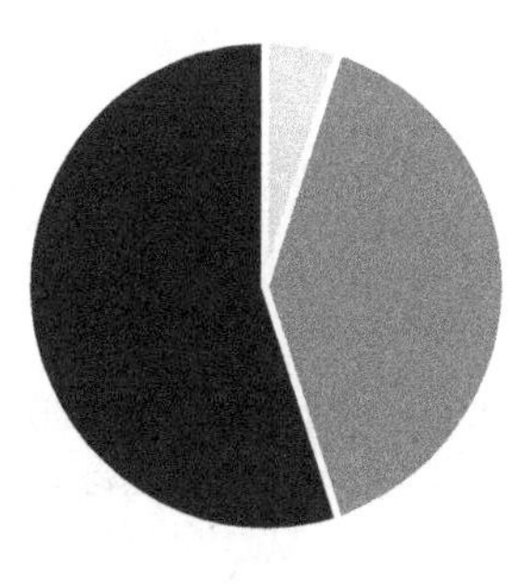

a. 5%
b. 25%
c. 40%
d. 55%

20. Which fraction represents the greatest part of the whole?

a. $\frac{1}{4}$

b. $\frac{1}{3}$

c. $\frac{1}{5}$

d. $\frac{1}{2}$

21. What other operation could be utilized to teach the process of dividing 9453 by 24 besides division?

a. Addition
b. Exponents
c. Multiplication
d. Subtraction

22. Divide, express with a remainder $188 \div 16$.

a. $1\frac{3}{4}$

b. $10\frac{3}{4}$

c. $11\frac{3}{4}$

d. $111\frac{3}{4}$

23. Which common denominator would be used to evaluate $\frac{2}{3} + \frac{4}{5}$?
    a. 3
    b. 5
    c. 10
    d. 15

24. What operation are students taught to repeat to evaluate an expression involving an exponent?
    a. Addition
    b. Division
    c. Multiplication
    d. Subtraction

25. The following table shows the temperature readings in Ohio during the month of January. How many more times was the temperature between 28-30 degrees than between 20-24 degrees?

Maximum Temperatures *in degrees*	**Tally marks**	**Frequency**
20 - 22	I	1
22 - 24	~~IIII~~ II	7
24 - 26	~~IIII~~	5
26 - 28	~~IIII~~ IIII	9
28 - 30	~~IIII~~ ~~IIII~~	10

    a. 4 times
    b. 5 times
    c. 9 times
    d. 10 times

26. Which of the following are units that would be taught in a lecture covering the metric system?
    a. Inches, feet, miles, pounds
    b. Teaspoons, tablespoons, ounces
    c. Kilograms, grams, kilometers, meters
    d. Millimeters, centimeters, meters, pounds

27. Which important mathematical property is shown in the expression: $(7 \times 3) \times 2 = 7 \times (3 \times 2)$?
    a. Associative property
    b. Commutative property
    c. Distributive property
    d. Multiplicative inverse

28. A grocery store is selling individual bottles of water, and each bottle contains 750 milliliters of water. If 12 bottles are purchased, what conversion will correctly determine how many liters that customer will take home?
    a. 10 liters equals 1 milliliter
    b. 100 milliliters equals 1 liter
    c. 1,000 liters equals 1 milliliter
    d. 1,000 milliliters equals 1 liter

29. If a student evaluated the expression $(3 + 7) - 6 \div 2$ to equal 2 on an exam, what error did she most likely make?
    a. There was no error. 2 is the correct answer.
    b. She did not perform the operation within the grouping symbol first.
    c. She divided first instead of the addition within the grouping symbol.
    d. She performed the operations from left to right instead of following order of operations.

30. Angie wants to shade $\frac{3}{5}$ of this strip. Which is the correct representation of $\frac{3}{5}$?

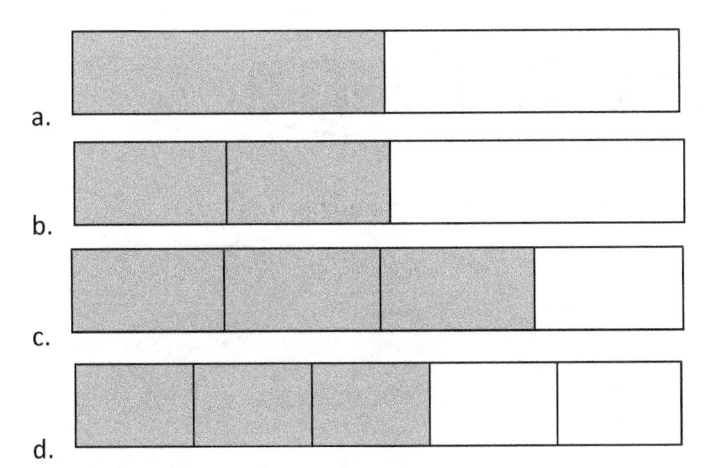

31. One digit in the following number is in **bold** and the other is underlined: 36,**6**01.
Which of the following statement about the underlined digit is true?
    a. Its value is $\frac{1}{10}$ the value of the bold digit.
    b. Its value is 10 times the value of the bold digit.
    c. Its value is 60 times the value of the bold digit.
    d. Its value is 100 times the value of the bold digit.

32. An angle measures 54 degrees. In order to correctly determine the measure of its complementary angle, what concept is necessary?
    a. Complementary angles are always acute.
    b. Complementary angles sum up to 360 degrees.
    c. Two complementary angles sum up to 90 degrees.
    d. Two complementary angles sum up to 180 degrees.

33. What tool could be used in the classroom to determine how many feet are in a yard?
    a. Compass
    b. Meter stick
    c. Ruler
    d. Yard stick

34. Ming would like to share his collection of 16 baseball cards with his three friends. He has decided that he will divide the collection equally among himself and his friends. Which of the following shows the correct grouping of Ming's cards?

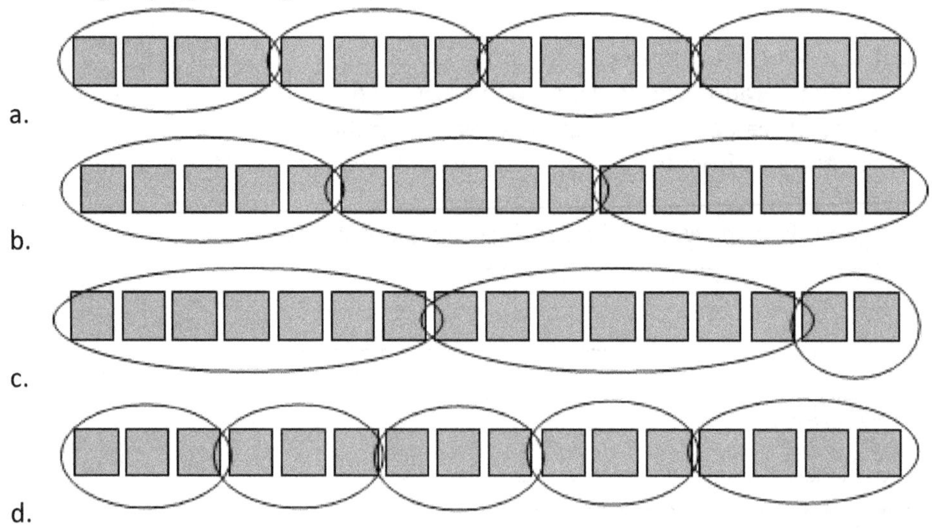

35. Which fractions are equivalent, or would fill the same portion on a number line?
    a. $\frac{2}{4}$ and $\frac{3}{8}$
    b. $\frac{1}{2}$ and $\frac{4}{8}$
    c. $\frac{3}{6}$ and $\frac{3}{5}$
    d. $\frac{2}{4}$ and $\frac{5}{8}$

36. Which of the following is an equivalent measurement for 1.3 cm?
    a. 0.13 m
    b. 0.013 m
    c. 0.13 mm
    d. 0.013 mm

37. Tommy was asked to round 84,789 to the nearest hundred. He came up with 85,000. He made a mistake. Choose the statement that explains what he did wrong.
    a. He should have used the 7 in the hundreds place to round to 85,789.
    b. He should have used the 7 in the hundreds place to round down to 84,000.
    c. He should have used the 8 in the tens place to round the 789 up to an 800, yielding 84,800.
    d. He should have used the 8 in the tens place to round the 789 down to a 700, yielding 84,700.

38. How would you write 5 as an ordinal number?
    a. fifd
    b. fifth
    c. five
    d. fivth

Reading Comprehension

Read the following passage, and then answer questions 1–5.

1 When renovating a home, there are several ways to save money. In order to keep a project cost
2 effective, "Do It Yourself," otherwise known as "DIY," projects help put money back into the
3 homeowner's pocket. For example, instead of hiring a contractor to do the demo, rent a
4 dumpster and do the demolition. Another way to keep a home renovation cost effective is to
5 compare prices for goods and services. Many contractors or distributors will match prices from
6 competitors. Finally, if renovating a kitchen or bathroom, leave the layout of the plumbing and
7 electrical the same. Once the process of moving pipes and wires is started, dollars start adding
8 up. Overall, home renovations can be a pricey investment, but there are many ways to keep
9 project costs down.

1. Which of the following statements is true based on the information in the passage?
    a. Home improvement projects can be expensive, but there are ways to keep costs down.
    b. Many contractors and distributors charge more than competitors for goods and services.
    c. Home renovations require a lot of work, which is why a contractor should be hired to complete the job.
    d. It is not necessary for homeowners to compare prices of contractors because they are their own best bet.

2. What is the meaning of the following sentence? "Do It Yourself," otherwise known as "DIY," projects help put money back into the homeowner's pocket.
    a. Homeowners get paid to do their own renovations.
    b. Homeowners save money by doing home repairs themselves.
    c. Hiring a contractor is more cost-effective than doing your own repairs.
    d. Homeowners will find money in their house while they are doing repairs.

3. Based on the opinion of the author, readers can infer that the author is likely which of the following?
    a. Someone who likes deals
    b. Someone who is very rich
    c. Someone who is a contractor
    d. Someone who is a distributor

4. Which of the following correctly lists the ways to keep renovation costs down, according to the author?
    a. Rent a dumpster, compare prices for goods and services, change the pipes and wires.
    b. Rent a dumpster, compare prices for goods and services, keep the layout of plumbing and electric.
    c. Hire a contractor for the demolition, compare prices for goods and services, change the pipes and wires.
    d. Hire a contractor for the demolition, compare prices for goods and services, keep the layout of plumbing and electric.

5. What is the meaning of *demo* as it is used in the passage?
    a. Democratic
    b. Demography
    c. Demolition
    d. Demonstration

Read the following poem, and then answer questions 6–10.

1 Standing in front of the mirror, I like to look at my face

2 I smile and frown and laugh and scream, emotions all over the place

3 Sometimes I stand between my mom and dad, all three of us in a row

4 We look each other up and down from the tops of our heads to the tips of our toes

5 My mom says I have her nose and her ears and a smile just like my dad

6 Our shirts and pants look different though because I wear jeans and dad wears plaid

7 Our hair color is also different though, which is confusing to me

8 Dad has black, mom has blond, but mine is brown like the bark of a tree

9 My teacher told me we inherit genes from our parents that affect how we look and act

10 Some of our features look like one or both of them, while some are unique to us in fact.

11 I am glad that I carry parts of mom and dad on my face and in my heart

12 That way they are with me wherever I go, even when we are apart.

6. Which of the following pairs of words in the poem are homophones?
    a. Fact and act
    b. Plaid and dad
    c. Genes and jeans
    d. Mirror and inherit

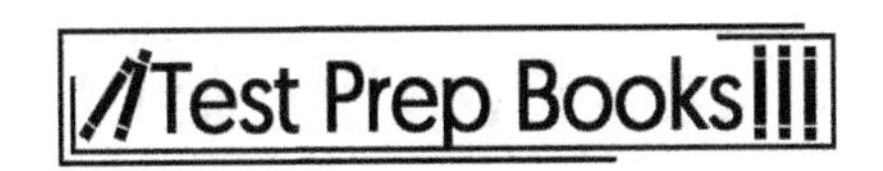

7. Which of the following lines most likely was meant to have a figurative, not literal, meaning?
   a. We look each other up and down.
   b. My mom says I have her nose and her ears.
   c. Standing in front of the mirror, I like to look at my face.
   d. My teacher told me we inherit genes from our parents that affect how we look and act.

8. What does this poem teach readers?
   a. Children and parents wear different types of pants.
   b. Children tend to look like people they are related to.
   c. You should carry a mirror with you wherever you go.
   d. If your hair color is brown, you don't look like your parents.

9. Which of the following is likely not an inherited trait?
   a. Behavior
   b. Genes
   c. Jeans
   d. Facial features

Read the following sentence, and answer the question below.

"Dad has black, mom has blond, but mine is brown like the bark of a tree"

10. In this sentence, the author is using what type of figurative language?
   a. Hyperbole
   b. Metaphor
   c. Simile
   d. Personification

The next five questions are based on the following passage.

1 George Washington emerged out of the American Revolution as an unlikely champion of liberty.
2 On June 14, 1775, the Second Continental Congress created the Continental Army, and John
3 Adams, serving in the Congress, nominated Washington to be its first commander. Washington
4 fought under the British during the French and Indian War, and his experience and prestige
5 proved instrumental to the American war effort. Washington provided invaluable leadership,
6 training, and strategy during the Revolutionary War. He emerged from the war as the
7 embodiment of liberty and freedom from tyranny.

8 After vanquishing the heavily favored British forces, Washington could have pronounced himself
9 as the autocratic leader of the former colonies without any opposition, but he famously refused
10 and returned to his Mount Vernon plantation. His restraint proved his commitment to the
11 fledgling state's republicanism. Washington was later unanimously elected as the first American
12 president. But it is Washington's farewell address that cemented his legacy as a visionary worthy
13 of study.

14 In 1796, President Washington issued his farewell address by public letter. Washington enlisted
15 his good friend, Alexander Hamilton, in drafting his most famous address. The letter expressed
16 Washington's faith in the Constitution and rule of law. He encouraged his fellow Americans to
17 put aside partisan differences and establish a national union. Washington warned Americans
18 against meddling in foreign affairs and entering military alliances. Additionally, he stated his
19 opposition to national political parties, which he considered partisan and counterproductive.

20 Americans would be wise to remember Washington's farewell, especially during presidential
21 elections when politics hits a fever pitch. They might want to question the political institutions
22 that were not planned by the Founding Fathers, such as the nomination process and political
23 parties themselves.

11. Which of the following statements is logically based on the information contained in the passage above?

a. George Washington would probably not approve of modern political parties.
b. George Washington would have opposed America's involvement in the Second World War.
c. George Washington's background as a wealthy landholder directly led to his faith in equality, liberty, and democracy.
d. George Washington would not have been able to write as great a farewell address without the assistance of Alexander Hamilton.

12. Which of the following statements is the best description of the author's purpose in writing this passage about George Washington?

c. To note that George Washington was more than a famous military hero
b. To introduce George Washington to readers as a historical figure worthy of study
d. To convince readers that George Washington is a hero of republicanism and liberty
a. To inform American voters about a Founding Father's sage advice on a contemporary issue and explain its applicability to modern times

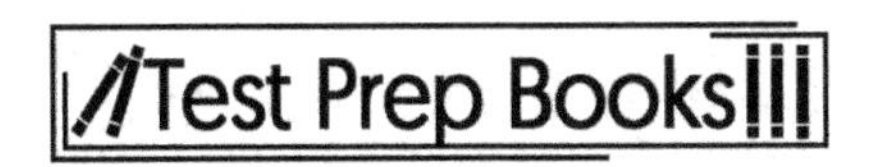

13. In which of the following materials would the author be the most likely to include this passage?
    a. An obituary
    b. A fictional story
    c. A history textbook
    d. A newspaper editorial

14. Which historical event happened before Washington was elected president?
    a. Washington issued his farewell address by public letter.
    b. Washington fought under the British during the French and Indian War.
    c. Washington enlisted the help of Alexander Hamilton in drafting his most famous address.
    d. Washington pronounced himself as the autocratic leader of the former colonies without opposition.

15. What is the meaning of the word *drafting* as it is used in the passage?
    a. Breeze
    b. Preparing
    c. Receipt
    d. Recruiting

Read the following passage, and then answer questions 16–20.

1 Horses are beautiful creatures. Anyone who has watched a horse race or a jumping competition
2 can attest to the fact that they are both powerful and graceful. Historically, horses have been
3 used for a variety of purposes since their domestication thousands of years ago. Horses played
4 an important role in history by providing means of transportation before automobiles. They also
5 were used for working purposes such as pulling a plow or being used to herd cattle. Horses have
6 also been used in warfare as far back as 3000 BC. Even today, horses are used on ranches and by
7 mounted police officers and other types of law enforcement officers. Horses are not only
8 beautiful but useful as well.

16. What is the main purpose of the passage?
    a. To describe the different breeds of horses
    b. To contrast horses with other forms of travel
    c. To inform the reader about horses and their uses
    d. To persuade the reader that horses are superior to other animals

17. What is the meaning of the word *domestication* in the passage?
    a. The evolution of horses
    b. The use of horses in warfare
    c. The taming of horses by humans
    d. The spread of horses throughout world regions

18. Which of the following is an opinion from the passage?
    a. Horses are beautiful creatures.
    b. Horses have also been used in warfare as far back as 3000 BC.
    c. Horses played an important role in history by providing means of transportation before automobiles.
    d. Even today, horses are used on ranches and by mounted police officers and other types of law enforcement officers.

19. Which of the following is presented in the passage as a use for horses historically?
    a. Horse races
    b. Herding cattle
    c. Pleasure riding
    d. Jumping competitions

20. Which of the following might the author use to describe horses?
    a. Grisly
    b. Majestic
    c. Obsolete
    d. Shabby

Read the following passage, and then answer questions 21–25.

1 Recently, I attended a marching band contest with my fellow classmates. We were incredibly
2 excited to get to perform in front of our parents and the judges. However, it seemed like maybe
3 it wasn't my lucky day. First, I fell getting my instrument out of the trailer and knocked over a
4 row of other instruments. They went down like dominoes. I was mortified. Then, I spilled soda
5 on the sleeve of my white coat on accident. Next, as we were getting ready to march onto the
6 field, I was getting my horn ready and accidentally punched my best friend in the nose. I told her
7 I was sorry, but I still felt horrible.

8 Finally, it was time to march. We held our instruments high and marched in a tight formation
9 onto the field. I completely forgot to worry about all the things that had happened that day. I
10 even managed to ace my solo. The judges loved us! I guess it turned out to be a good day after
11 all.

21. Which of the following best expresses the main idea of the passage?
    a. Always try your best.
    b. Overcome your fears.
    c. Take the good with the bad.
    d. Hard work creates the best results.

22. Which of the following is an example of figurative language in the passage?
    a. I even managed to ace my solo.
    b. They went down like dominoes.
    c. I told her I was sorry, but I still felt horrible.
    d. We held our instruments high and marched in a tight formation onto the field.

23. In the passage, what is the meaning of the word "mortified"?
    a. Delighted
    b. Embarrassed
    c. Exhausted
    d. Scared

24. What line from the passage contributes to the idea that the day turned out to be a good day after all?
   a. The judges loved us!
   b. Finally, it was time to march onto the field.
   c. Then, I spilled soda on the sleeve of my white coat on accident.
   d. We were incredibly excited to get to perform in front of our parents and the judges.

25. Which happened first in the paragraph?
   a. The band marched onto the field.
   b. The narrator fell out of the trailer.
   c. The row of instruments was knocked over.
   d. The narrator punched a friend in the nose.

Math Achievement

1. What is the 42nd item in the pattern: ▲○○□▲○○□▲...?
   a. ○
   b. ▲
   c. □
   d. None of the above

2. Which of the following is NOT a way to write 40 percent of *N*?
   a. $(0.4)N$
   b. $\frac{2}{5}N$
   c. $40N$
   d. $\frac{4N}{10}$

3. Which is closest to 17.8×9.9?
   a. 140
   b. 180
   c. 200
   d. 350

4. Add $5{,}089 + 10{,}323$
   a. 5,234
   b. 15,234
   c. 15,402
   d. 15,412

5. A closet is filled with red, blue, and green shirts. If there are 54 shirts total, and 12 are green and 28 are red, which equation below could be used to calculate the number of blue shirts?
   a. $54 - 12 - 28 = \square$
   b. $54 = 28 - 12 + \square$
   c. $54 - \square = 28 + 12$
   d. $54 + 12 + 28 = \square$

6. Add and express in reduced form $\frac{5}{12} + \frac{4}{9}$

a. $\frac{1}{3}$

b. $\frac{9}{17}$

c. $\frac{3}{5}$

d. $\frac{31}{36}$

7. Apples cost $2 each, while oranges cost $3 each. Maria purchased 10 fruits in total and spent $22. How many apples did she buy?

a. 5
b. 6
c. 7
d. 8

8. Subtract $9{,}576 - 891$.

a. 8,325
b. 8,685
c. 9,685
d. 10,467

9. Determine the next number in the following series: $1, 3, 6, 10, 15, 21, \ldots$

a. 26
b. 27
c. 28
d. 29

10. Subtract $50.888 - 13.091$.

a. 33.817
b. 37.797
c. 37.979
d. 63.979

11. Mom's car drove 72 miles in 90 minutes. There are 5280 feet per mile. How fast did she drive in feet per second?

a. 0.009 feet per second
b. 0.8 feet per second
c. 48.9 feet per second
d. 70.4 feet per second

12. Subtract and express in reduced form $\frac{23}{24} - \frac{1}{6}$.

a. $\frac{19}{24}$

b. $\frac{4}{5}$

c. $\frac{22}{18}$

d. $\frac{11}{9}$

13. What is the volume of a cube with the side equal to 3 inches? ($V = s^3$ where V is volume and s is the length of one side)

a. 3 in^3
b. 6 in^3
c. 9 in^3
d. 27 in^3

14. What is the expression that represents three times the sum of twice a number and one minus 6?

a. $2x + 1 - 6$
b. $3x + 1 - 6$
c. $3(x + 1) - 6$
d. $3(2x + 1) - 6$

15. Express as an improper fraction $11\frac{5}{8}$.

a. $\frac{16}{11}$

b. $\frac{19}{5}$

c. $\frac{55}{8}$

d. $\frac{93}{8}$

16. 6 is 30% of what number?

a. 18
b. 20
c. 24
d. 26

17. $4\frac{1}{3} + 3\frac{3}{4} =$

a. $6\frac{5}{12}$

b. $7\frac{7}{12}$

c. $8\frac{1}{12}$

d. $8\frac{2}{3}$

18. A square has a side length of 4 inches. A triangle has a base of 2 inches and a height of 8 inches. What is the total area of the square and triangle?
    a. 24 square inches
    b. 28 square inches
    c. 32 square inches
    d. 36 square inches

19. Which number is a multiple of both 6 and 8?
    a. 30
    b. 48
    c. 180
    d. 440

20. What is the perimeter of the following figure?

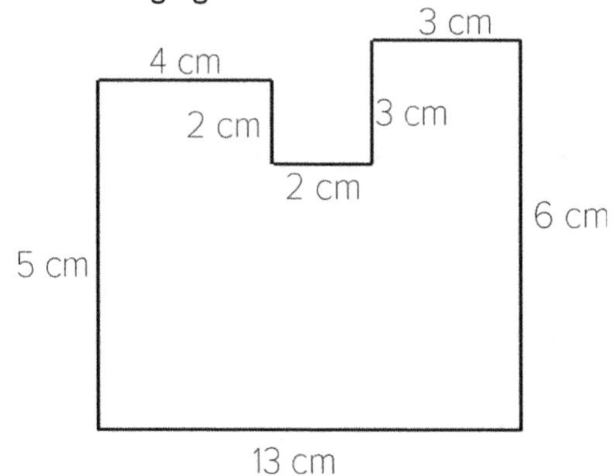

    a. 35
    b. 38
    c. 40
    d. 65

21. What is the standard form for four hundred fifty-six thousand one hundred eighty-three?
    a. 450,183
    b. 451,830
    c. 456, 083
    d. 456,183

22. Four students measured the height of their individual pea plants each week for four weeks. The results are shown below.

	Week 1	Week 2	Week 3	Week 4
Chris	1 in.	1.5 in.	1.7 in.	2.1 in.
David	.9 in.	1.25 in.	1.6 in.	1.9 in.
Emily	1.1 in.	1.25 in.	1.4 in.	2 in.
Lauren	1 in.	1.3 in.	1.75 in.	2.25 in.

Which student's plant grew the most from Week 3 to Week 4?
- a. Chris
- b. David
- c. Emily
- d. Lauren

23. In May of 2010, a couple purchased a house for $100,000. In September of 2016, the couple sold the house for $93,000 so they could purchase a bigger one to start a family. How many months did they own the house?
- a. 54
- b. 76
- c. 85
- d. 93

24. Mo needs to buy enough material to cover the walls around the stage for a theater performance. If he needs 79 feet of wall covering, what is the minimum number of yards of material he should purchase if the material is sold only by whole yards?
- a. 23 yards
- b. 25 yards
- c. 26 yards
- d. 27 yards

25. The value of 6 x 12 is the same as:
- a. 7 x 4 x 3
- b. 6 x 6 x 3
- c. 2 x 4 x 4 x 2
- d. 3 x 3 x 4 x 2

26. The total perimeter of a rectangle is 36 cm. If the length is 12 cm, what is the width?
- a. 3 cm
- b. 6 cm
- c. 8 cm
- d. 12 cm

27. What's the probability of rolling a 6 at least once in two rolls of a die?
   a. $\frac{1}{36}$
   b. $\frac{1}{3}$
   c. $\frac{1}{6}$
   d. $\frac{11}{36}$

28. What is the value of *b* in this equation?

$$5b - 4 = 2b + 17$$

   a. 7
   b. 13
   c. 21
   d. 24

29. Dwayne has received the following scores on his math tests: 78, 92, 83, and 97. What score must Dwayne get on his next math test to have an overall average of 90?
   a. 89
   b. 95
   c. 98
   d. 100

30. What is the overall median of Dwayne's current scores: 78, 92, 83, 97?
   a. 19
   b. 83
   c. 85
   d. 87.5

Essay

Select a topic from the list below and write an essay. You may organize your essay on another sheet of paper.

Topic 1: Who is someone you want to be like and why?

Topic 2: If you could spend a day doing whatever you wanted in your hometown, what would it be and why?

Topic 3: Who is a celebrity you would love to meet? Why?

Answer Explanations #2

Verbal Reasoning

Synonyms

1. A: *Initiate* means to begin, start, or commence. Choice *B*, sedate, means to soothe or calm. Choice *C*, surmise, means to guess or wonder. Choice *D*, vacillate, means to sway back and forth.

2. B: *Passive* is most closely related to the word *indifferent*. Choice *A*, compatible, means to be agreeable or to get along with someone easily. Choice *C*, snub, is close to the word *passive*, but it's not as close in meaning as *indifferent*. *Snub* means to give someone the cold shoulder or to ignore someone purposefully. Choice *D*, *tranquil*, means calm or peaceful.

3. B: *Robust* means strong or sturdy. Choice *A*, abort, means to give something up that's unfinished. Choice *C*, verbose, means wordy or talkative. Choice *D*, weak, is the opposite of *robust*.

4. C: *Pragmatic* means sensible or down to earth. Choice *A*, fickle, means not reliable or dependable. Choice *B*, indignant, means to be outraged at some kind of injustice. Choice *D*, strenuous, is something that requires hard work or is difficult.

5. A: *Aloof* means remote or detached. Choice *B*, envious, means to be jealous. Choice *C*, jubilant, means to be joyful or happy. Choice *D*, limber, means to be flexible.

6. B: *Juvenile* means to be immature or young. Choice *A*, fatuous, means to be silly or foolish. Choice *C*, impoverished, means to be poor. Choice *D*, stable, means to be constant or fixed.

7. D: *Courteous* is most closely related to the word *polite*. Choice *A*, extravagant, means excessive or indulgent. Choice *B*, facile, means easy. Choice *D*, meager, means small or insufficient.

8. C: *Matrimony* and *marriage* are synonyms. Choice *A*, entity, means an object that exists as a particular unit. Choice *B*, hiatus, means an interruption or break. Choice *D*, parody, is an imitation or spoof.

9. B: *Pacify* means to soothe or ease. Choice *A*, patent, means apparent or clear. Choice *C*, stagnate, means to deteriorate or fester. Choice *D*, trek, means to journey or hike.

10. A: To be *prudent* means to be careful or cautious. Choice *B*, impatient, means unable to wait. Choice *C*, likeable, means friendly or kind. Choice *D*, tranquil, means to be quiet or peaceful.

11. B: *Colleague* means fellow coworker. Choice *A*, college, is spelled like *colleague*, but it does not mean the same thing. Choice *C*, executive, is someone in a leadership position. Choice *D*, subordinate, is someone who works under a position of authority.

12. C: *Essential* means necessary. Choice *A*, belligerent, means aggressive or argumentative. Choice *B*, mundane, means ordinary or banal. Choice *D*, pervasive, means to spread extensively.

13. C: *Orator* and *speaker* are synonyms. Choices *A*, *B*, and *D*, dancer, singer, and writer, are not considered orators.

14. D: To *scorn* means to ridicule or jeer. Choices *A*, *B*, and *C*, cheer, esteem, and praise are all antonyms, or opposite in meaning, of the word *scorn*.

15. B: *Cherish* means to adore or care for deeply. Choice *A*, admonish, means to advise or warn. Choice *C*, bombard, means to assault or attack. Choice *D*, command, means to direct or instruct.

16. A: *Astound* means to amaze. Choice *B*, espy, means to catch sight of. Choice *C*, promote, means to advance or help. Choice *D*, regard, means to give attention to something.

17. B: The word *lucrative* means productive. Choice *A*, apathetic, means lazy or inactive. Choice *C*, singular, means odd or unique. Choice *D*, supreme, means greatest or absolute.

Sentence Completion

18. D: The word *void* describes nothingness or emptiness. The choices *around* and *diffuse* are not part of the context; *diffuse* means to spread out. Space is not a planet, although plants are within space.

19. C: *Severe* reflects high intensity or very great. Cut and shake are verbs with actions that do not reflect severe. While something can be severely foul, foul is a broad description of something bad, but not necessarily of a high level of intensity.

20. D: *Verbose* reflects the tendency to talk a lot. Cranky has nothing to do with the learning to speak, and silent and quiet are both opposites.

21. A: To *disparage* someone is to put them down. Fixed, praised, and uplifted are all positive words meaning to support or encourage someone, so these are incorrect.

22. D: Looking at the root word ego, egotistic must have something to do with the self—in this case, excessive self-interest. Such a person tends to be the opposite of altruistic, which means selfless. The word amicable means friendly, so this choice is incorrect. Agrarian is an unrelated word concerning fields or farm lifestyle.

23. D: *Terrain*, from the Latin *terra*, refers to the earth or physical landscape. Arboreal describes things relating to trees. Celestial and firmament both describe the sky.

24. C: *Encompass* means to hold within or surround. Avoid and retreat are opposites. Brazen does not make sense in the context of a sentence; we are looking for a strategy that is used to defeat an army.

25. A: *Contrast* means to go against or to have a different perspective. Revive and improve do not fit in the sentence, in a grammatical or contextual sense. Partisan can mean favoring one side, but contrast is the best choice.

26. B: The Latin root, *asper*, means rough. *Exasperate*, then, means to make relations with someone rough or to rub them the wrong way. Breathe does not fit into the sentence. Humor is incorrect because the babysitter would not be stressed out if she experienced humor from Nathaniel. Stifle is close, but *exasperate* is a better fit due to the stress of the babysitter.

27. D: From the Latin root *liber*, meaning free, *liberate* means to free or release. Agitate means to annoy, which is not the correct meaning for this sentence. Fracture is to crack or break something. Instigate can mean to start.

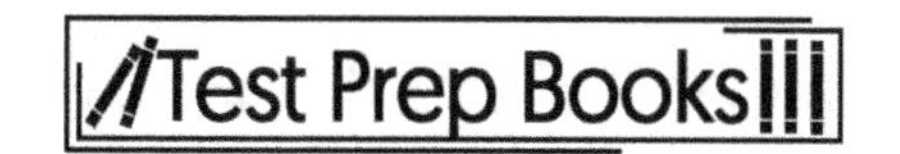

28. A: *Dejectedly* means depressed and sad. Jubilantly means happy, so it is the opposite of dejected. Passively means showing little or no response which is unlikely to be how the little girl would respond. Petrified means scared, so it wouldn't work in this sentence

29. A: *Barren* means deserted, void, lifeless, or having little. A desert is barren because it produces little vegetation. Fruitful, lavish, and sumptuous express richness and abundance, which contradict barren and the disappointment of a starving man.

30. A: *Chastise* means to reprimand severely. Choose, honor, and locate are incorrect in this situation.

31. B: *Forgive* literally means to pardon of sins or offenses. The boy would most likely forgive the girl whose apology was sincere.

32. A: *Altitude* is the measure of height, which is what we are looking for in this sentence. The other choices refer to attitude, a word spelled and pronounced similarly to altitude. Someone's attitude can reflect their behavior, reflect haughtiness, and influence their outlook. None of these terms describe altitude, which is the measure of height.

33. C: *Irate* comes from the Latin *ira-*, which gives it the meaning of angry or irritable. Amiable means happy and friendly. Innocent and tangible do not fit in this sentence because we are told the girl was "fuming inside."

34. C: *Provoke* means to incite a reaction in someone or something. Damage, inspire, and soothe do not work in this context.

Quantitative Reasoning

1. A: The tenths place is the first place to the right of the decimal point, so in the number 245.867, there is an 8 in the tenths place. The place value to the right of the tenths place, which would be the hundredths place, is what gets utilized. The value in the hundredths place is 6. The number in the place value to its right is greater than 4, so the 8 gets bumped up to 9. Everything to its right of the tenths place gets dropped, which means the result is 245.9.

2. C: Using the order of operations, multiplication and division are computed first from left to right. Multiplication is on the left; therefore, the teacher should perform multiplication first.

3. D: A factor of 36 is any number that can be divided into 36 and have no remainder. $36 = 36 \times 1, 18 \times 2, 9 \times 4, \text{and } 6 \times 6$. Therefore, it has 7 unique factors: 36, 18, 9, 6, 4, 2, and 1.

4. A: The number line is divided into ten portions, so each mark represents 0.1. Halfway between the 6th and 7th marks would be 0.65. Choice *B* shows 0.55, Choice *C* shows 0.25, and Choice *D* shows 0.45.

5. A: This choice can be determined by comparing the place values, beginning with that which is the farthest left; hundred-thousands, then ten-thousands, then thousands, then hundreds. It is in the hundreds place that Choice A is larger. Choice B is smaller in the ten-thousands place, Choice C is smaller in the tens place, and Choice D is smaller in the ten-thousands place.

6. B: This is the only shape that has no parallel sides, and therefore cannot be a parallelogram. Choice *A* has one set of parallel sides and Choices *C* and *D* have two sets of parallel sides.

7. A: The light gray portion represents $\frac{4}{5}$ and the dark gray portion represents $\frac{3}{5}$, to total $\frac{7}{5}$. Choice *B* is not correct because it misrepresents the dark gray portion. Choice *C* is not correct because it misrepresents the light gray portion. Choice *D* is not correct because it includes the 1 with the dark gray portion.

8. D: This answer is the only one that carries out proper multiplication to get the correct result (as seen below).

$$\begin{array}{r} 2^{1}6 \\ \times\underline{1\ 2} \\ 5\ 2 \\ +\underline{2^{1}6\ 0} \\ 3\ 1\ 2 \end{array}$$

9. C: To add fractions, the denominator must be the same. This is the only choice with both denominators of 6. Adding the numerators totals 7, for a fraction of $\frac{7}{6}$. Choice *A* equals $\frac{7}{3}$, Choice *B* equals $\frac{7}{5}$, and Choice *D* equals $\frac{8}{4}$ or 2.

10. A: Angle *B* is an acute angle because it is smaller than a right angle, which is 90°. Therefore, we can immediately eliminate Choices *C* and *D*. To determine the measure of the angle, look at where the ray that is not along the bottom crosses the arc of the protractor. It falls between the 40 and 50; specifically, it is at 47°. If the ray along the bottom was going towards the left and the other ray stayed where it is now, the angle would be obtuse, and the number of degrees would be read as 133°.

11. D: Numbers should be lined up by decimal places before subtraction is performed. This is because subtraction is performed within each place value. The other operations, such as multiplication, division, and exponents (which is a form of multiplication), involve ignoring the decimal places at first and then including them at the end.

12. B: If 60% of 50 workers are women, then there are 30 women working in the office. If half of them are wearing skirts, then that means 15 women wear skirts. Since none of the men wear skirts, this means there are 15 people wearing skirts.

13. A: $\frac{2}{9}$

Set up the problem and find a common denominator for both fractions.

$$\frac{43}{45} - \frac{11}{15}$$

Multiply each fraction across by 1 to convert to a common denominator.

$$\frac{43}{45} \times \frac{1}{1} - \frac{11}{15} \times \frac{3}{3}$$

Once over the same denominator, subtract across the top.

$$\frac{43 - 33}{45} = \frac{10}{45}$$

Reduce.

$$\frac{10 \div 5}{45 \div 5} = \frac{2}{9}$$

14. D: The volume of a cube is the length of the side cubed, and 5 centimeters cubed is 125 cm^3. Choice *A* is not the correct answer because that is 2×5 centimeters. Choice *B* is not the correct answer because that is 3×5 centimeters. Choice *C* is not the correct answer because that is 5×10 centimeters.

15. B: The number line is divided into 10 sections, so each portion represents 0.1. Because the number line begins at 1 and ends at 2, the number in question would be between those two numbers. Since there are only two portions out of ten marked, this represents the number 1.2. All other choices are incorrect due to a misreading of the number line.

16. D: Area = length x width. The answer must be in square inches, so all values must be converted to inches. $\frac{1}{2}$ ft is equal to 6 inches. Therefore, the area of the rectangle is equal to $6 \times \frac{11}{2} = \frac{66}{2} = 33$ square inches.

17. C: The teacher would be introducing fractions. If a pie was cut into 6 pieces, each piece would represent $\frac{1}{6}$ of the pie. If one piece was taken away, $\frac{5}{6}$ of the pie would be left over.

18. B: If an array were to be used, 123 items could be divided up into 4 groups of 30, with 3 left over. Choices *A*, *C*, and *D* are misrepresentations of the correct grouping and not equal to 30 with a remainder of 3.

19. C: The sum total percentage of a pie chart must equal 100%. Since the CD sales take up less than half of the chart and more than a quarter (25%), it can be determined to be 40% overall. This can also be measured with a protractor. The angle of a circle is 360°. Since 25% of 360° would be 90° and 50% would be 180°, the angle percentage of CD sales falls in between; therefore, it would be Choice *C*.

20. D: Even though all of the fractions have the same numerator, this is the one that represents the greatest part of the whole. All other choices are smaller portions of the whole, as seen by this graphic representation.

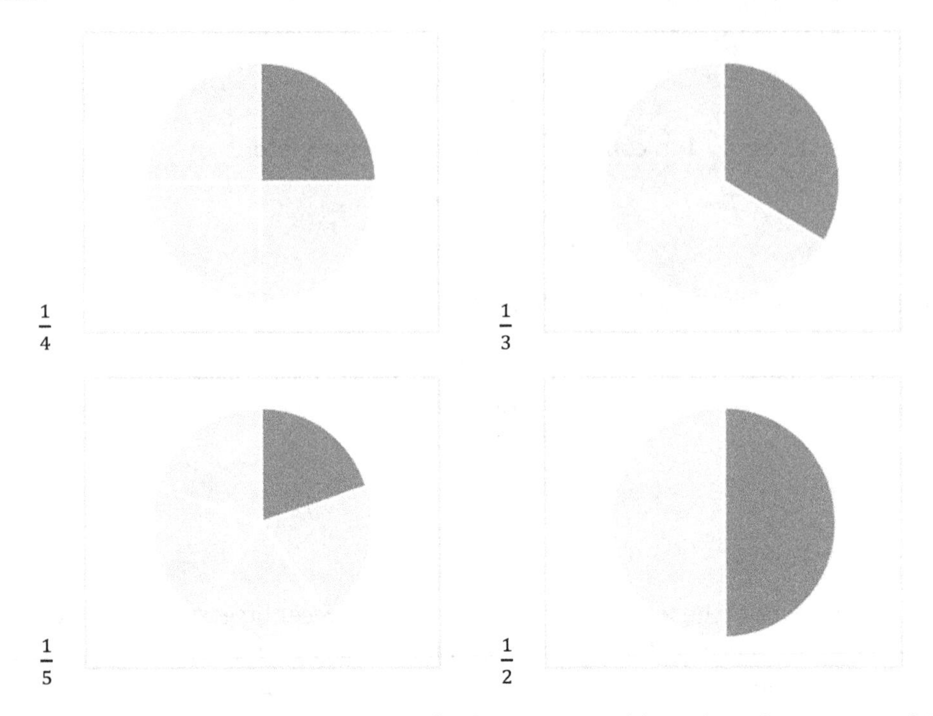

21. D: Division can be computed as a repetition of subtraction problems by subtracting multiples of 24.

22. C: $11\frac{3}{4}$

Set up the division problem.

$$16\overline{)188}$$

16 does not go into 1 but does go into 18 so start there.

$$\begin{array}{r} 11 \\ 16\overline{)188} \\ -16 \\ \hline 28 \\ -16 \\ \hline 12 \end{array}$$

The result is $11\frac{12}{16}$

Reduce the fraction for the final answer: $11\frac{3}{4}$

23. D: A common denominator must be found. The least common denominator is 15 because it has both 5 and 3 as factors. The fractions must be rewritten using 15 as the denominator.

24. C: A number raised to an exponent is a compressed form of multiplication. For example:

$$10^3 = 10 \times 10 \times 10$$

25. A: To calculate this, the following equation is used: $10 - (5 + 1) = 4$. The number of times the temperature was between 36-38 degrees was 10. Finding the total number of times the temperature was between 28-32 degrees requires totaling the categories of 28-30 degrees and 30-32 degrees, which is $5 + 1 = 6$. This total is then subtracted from the other category in order to find the difference. Choice *B* only subtracts the 30-32 degrees category from the 36-38 degrees category. Choice *C* only subtracts the 28-30 degrees from the 36-38 degrees category. Choice *D* is simply the number from the 36-38 degrees category.

26. C: Inches, pounds, and baking measurements, such as tablespoons, are not part of the metric system. Kilograms, grams, kilometers, and meters are part of the metric system.

27. A: It shows the associative property of multiplication. The order of multiplication does not matter, and the grouping symbols do not change the final result once the expression is evaluated.

28. D: $12 \times 750 = 9{,}000$. Therefore, there are 9,000 milliliters of water, which must be converted to liters. 1,000 milliliters equals 1 liter; therefore, 9 liters of water are purchased.

29. D: According to order of operations, the operation within the parentheses must be completed first. Next, division is completed and then subtraction. Therefore, the expression is evaluated as:

$$(3 + 7) - 6 \div 2$$
$$10 - 6 \div 2 = 10 - 3$$
$$7$$

In order to incorrectly obtain 2 as the answer, the operations would have been performed from left to right, instead of following PEMDAS.

30. D: This solution shows the strip separated into 5 pieces, which is necessary for it to be filled in to show $\frac{3}{5}$. Choice *A* shows the strip filled to $\frac{1}{2}$, and Choice *B* shows the strip filled to $\frac{2}{4}$, which is also $\frac{1}{2}$. Neither of these selections is correct. While Choice *C* shows 3 portions filled, the total number of portions is only 4, making the fraction filled $\frac{3}{4}$. This is also an incorrect choice.

31. B: The underlined digit is the 6 in 6,000. The bold digit is the 6 in 600. Because 6,000 is equal to 6000×10, we know that the underlined 6 has a value that is 10 times that of the bold 6. Additionally, the base-10 system we use helps us determine that the place value increases by a multiple of ten when you go from the right to the left.

32. C: The measure of two complementary angles sums up to 90 degrees. $90 - 54 = 36$. Therefore, the complementary angle is 36 degrees.

33. D: A yard stick is a tool that can be used to measure how many feet are in a yard. Simply take the yard stick and count out sections of 12 inches (12 inches is equal to one foot) until you run out of stick. You will find that there are 3 feet in a yard.

34. A: This choice shows that Ming plus his three friends (1 + 3 = 4) is the number of divisions necessary to split the lot of cards evenly (16 ÷ 4 = 4). There would need to be four groups of 4 cards each, or $\frac{4}{16}$, which is $\frac{1}{4}$ of the total cards. The other choices do not correctly divide the cards into even groupings.

35. B: $\frac{1}{2}$ is the same fraction as $\frac{4}{8}$, and would both fill up the same portion of a number line.

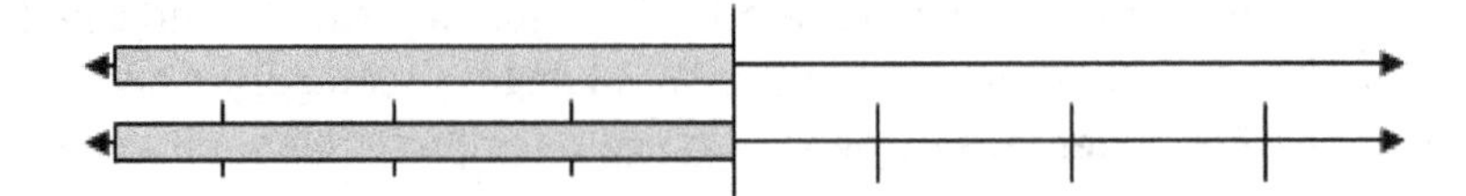

36. B: 100 cm is equal to 1 m. 1.3 divided by 100 is 0.013. Therefore, 1.3 cm is equal to 0.013 m. Because 1 cm is equal to 10 mm, 1.3 cm is equal to 13 mm.

37. C: Tommy should have used the 8 in the tens place to round the 789 up to an 800, yielding 84,800. The hundred place value is located three digits to the left of the decimal point (the digit 7). The digit to the right of the place value is examined to decide whether to round up or keep the digit. In this case, the digit 8 is 5 or greater so the hundred place is rounded up. When rounding up, any digits to the left of the place value being rounded remain the same and any to its right are replaced with zeros. Therefore, the number is rounded to 84,800. Tommy accidentally used the digit in the hundreds place to round to the nearest thousand, instead of using the digit in the tens place to round to the nearest hundred.

38. B: The correct answer is fifth. It is the fifth ordinal number.

Reading Comprehension

1. A: The main idea of the passage is that home improvement projects can be expensive. However, there are ways to keep the costs down. The details of the other choices go against what the passage says. So, they are incorrect. For example, the passage says that contractors will often price-match competitors. This makes Choice *B* incorrect. Choice *C* is incorrect because one of the author's main points is that they do not need to hire a contractor for all renovations. They can do it themselves as a DIY project. Choice *D* is incorrect because the passage does mention that some projects require professionals, but that comparing prices can minimize costs.

2. B: Something that is "like money on your pocket" means that it is money savings or a deal. This is a common phrase. Choice *A* doesn't make sense, because how would they get paid to do their own repairs? Who would pay them and why? Choice *C* is incorrect based on the main idea and details mentioned in the passage. For example, readers are encouraged in the passage not to get a contractor to do the demolition, but to rent a dumpster themselves to save money. While some change may be found in couch cushions or loose money around some people's homes, Choice *D*, finding money while doing repairs, is not a likely choice.

3. A: Readers can guess that the author of the passage likes to find deals. For this reason, Choice *B* is more unlikely than Choice *A*, because if the narrator was very rich, they may be less interested in strategies to save money. The advice in the passage is mostly doing projects yourself, so the narrator is probably not a contractor or distributor. For this reason, Choices *C* and *D* are incorrect.

4. B: Readers must look carefully over the paragraph to find the author's advice. Readers will find that the author of the passage says to rent a dumpster, compare prices for goods and services, and keep the layout of plumbing and electric. The other answer choices had at least one incorrect suggestion.

5. C: Although demo can be used as a shortened version of all these choices, it is used in this passage in reference to demolition. You can find its definition in the sentence "For example, instead of hiring a contractor to do the demo, rent a dumpster and do the demolition."

6. C: Homophones are words that have the same pronunciation but different meanings or spellings. A good example of this is *new* and *knew*. New means something that has only existed for a short period of time. Knew, on the other hand, refers to the past tense of the verb to know, meaning to be aware of something. To answer this question correctly, test takers must be able to properly read the words genes and jeans. Genes are the genetic material that dictate a person's traits. Jeans are denim pants. The other answer choices provided were either pairs of rhyming words or simply two unrelated words.

7. B: My mom says I have her nose and her ears is an example of figurative language. Figurative language describes things using creative and imaginative terms, similes and metaphors, and poetic language. The words do not mean exactly what they say. In this case, the child narrating the poem does not truly have the mom's nose and ears. Those features are stuck on the mom's face! Instead, what is meant is that the child's features look very much like the nose and ears of the mom. This is a poetic expression. This is not a fact that should be taken with a literal meaning. The other choices are examples of lines in the poem with more literal meanings. Literal is when words mean what they say without any sort of imaginative language.

8. B: This poem teaches readers that children often have characteristics that look like their parents. This is because we inherit genes from our parents and these genes have something called DNA, which tells our body what we should look like. The other choices are not really lessons that the poem is teaching readers. Children and parents may wear different types of pants, but they don't have to. They can wear pants that are very similar. It is not usually necessary to carry a mirror wherever one goes. Lastly, some children with brown hair look like their parents who may also have brown hair.

9. C: Genes are inherited, but jeans are clothing. Blue jeans may be passed down from an older sibling or parent as a hand-me-down. However, they are not a characteristic of an individual. Traits are what a person is as a whole.

10. C: The author is comparing the color of their hair to the color of tree bark, so it is either a metaphor or simile. Since the word "like" is used, this makes it a simile, so Choice *C* is correct. Choice *A*, hyperbole, is an unrealistic exaggeration and is incorrect. Choice *B*, metaphor, is incorrect because metaphors do not use "like" or "as" to compare two things. Choice *D*, personification, gives humanlike qualities to inanimate objects. A human and a tree are mentioned in this sentence, but the tree is not behaving like a human, so this choice is incorrect.

11. A: Washington's warning against meddling in foreign affairs does not mean that he would oppose wars of every kind, so Choice *B* is wrong. Although Washington was from a wealthy background, the passage does not say that his wealth led to his republican ideals, so Choice *C* is not supported. Choice *D* is also unjustified since the author does not indicate that Alexander Hamilton's assistance was absolutely necessary. Choice *A* is correct because the farewell address clearly opposes political parties and partisanship. The author then notes that presidential elections often hit a fever pitch of partisanship. Thus, it follows that George Washington would not approve of modern political parties and their involvement in presidential elections.

12. D: The author finishes the passage by applying Washington's farewell address to modern politics, so the purpose probably includes this application. Choice *A* is wrong because the author is not fighting a common perception that Washington was merely a military hero. Choice *B* is wrong because George Washington is already a well-established historical figure; furthermore, the passage does not seek to introduce him. Choice *C* is wrong because the author is not convincing readers. Persuasion does not correspond to the passage. Choice *D* states the primary purpose.

13. D: Choice *A* is false because the piece is not a notice or announcement of Washington's death. Choice *B* is false because it is not fiction, but a historical writing. Choice *C* is wrong because the last paragraph is not appropriate for a history textbook. Choice *D* is correct. The passage is most likely to appear in a newspaper editorial because it cites information relevant and applicable to the present day, a popular format in editorials.

14. B: Choice *B* is correct because the French and Indian War took place before Washington was elected president. His contribution in the French and Indian War and Revolutionary War led to his election. Because this passage is in chronological order, we can eliminate Choices *A* and *C*, because both of these take place after Washington was elected president. Choice *D* is incorrect because Washington could have done this, but refused to do so and returned to his plantation. Therefore, this is not a historical event.

15. B: Preparing. While all of these choices are synonyms of drafting, Choice *B* is correct because Alexander Hamilton helped Washington prepare his farewell address. Choice *A*, breeze, is incorrect, because wind was not used to write the farewell address. Choice *C*, receipt, is incorrect because Hamilton and Washington did not receive the farewell address. They wrote and sent it to others. Choice *D*, recruiting, is also incorrect because Hamilton and Washington did not choose or select anyone else to help them write the address according to the passage.

16. C: The main purpose of this passage is to inform the reader about horses and their uses. The passage does not discuss any particular breeds; therefore, Choice *A* is incorrect. Choice *B* is not correct because while automobiles are mentioned, they are not used as part of the main purpose of the passage. There is no mention of other animals, so Choice *D* is also incorrect.

17. C: "Domestication" refers to when an animal is tamed so that it can be kept by humans and trained to respond to its owner's commands. It does not refer to evolution, spread, or the use of horses in warfare.

18. A: "Horses are beautiful" expresses the author's opinion about the appearance of horses. The other choices are all facts that can be validated using outside sources.

19. B: Herding cattle is mentioned as one way that horses have been utilized in history. Horse races and jumping competitions are mentioned, but the author does not specify that these are historical uses of horses. Pleasure riding is not mentioned in the passage at all.

20. B: Majestic means impressive and great, which is similar to the author referring to horses as "beautiful" and "graceful". Grisly and shabby mean the opposite of majestic, so Choices *A* and *D* are incorrect. Choice *C*, obsolete, is incorrect, because the author mentions horses are still useful even today.

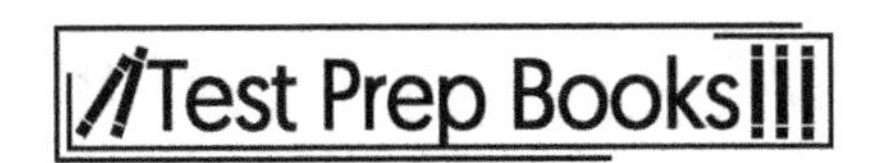

21. C: The day began with a series of unfortunate events, but in the end, it was a great day so, Choice *C* makes the most sense in the context of the passage. The narrator seemed to try his best and work hard but that was not the main idea of the passage. The narrator's fears were not mentioned in the passage.

22. B: "They went down like dominoes" is a simile, so this is an example of figurative language. The other choices do not represent figurative language.

23. B: Mortified means embarrassed or ashamed. After falling out of the trailer, the narrator is embarrassed about what has happened.

24. A: The fact that the band did well at the competition is the main reason the day turned out well, and this is supported by the approval of the judges. The act of marching onto the field was not enough to turn the day around. When the narrator spilled soda on his sleeve, this was part of what made him feel unlucky. The students' excitement was not enough to make the day great since there were several other incidents after that which marred the day.

25. B: The narrator fell out of the trailer first and then knocked the row of instruments over. Later in the passage, his friend gets punched in the nose. The band marching onto the field is one of the last things to happen in the passage.

Math Achievement

1. A: ○. The core of the pattern consists of 4 items: ▲○○□. Therefore, the core repeats in multiples of 4, with the pattern starting over on the next step. The closest multiple of 4 to 42 is 40. Step 40 is the end of the core (□), so step 41 will start the core over (▲) and step 42 is ○.

2. C: 40*N* would be 4000% of *N*. It's possible to check that each of the others is actually 40% of *N*.

3. B: Instead of multiplying these out, the product can be estimated by using $18 \times 10 = 180$. The error here should be lower than 15, since it is rounded to the nearest integer, and the numbers add to something less than 30.

4. D: 15,412

Set up the problem and add each column, starting on the far right (ones). Add, carrying anything over 9 into the next column to the left. Solve from right to left.

5. A: The sum of the shirts is 54. So, we know that the number of green shirts, plus the number of red shirts, plus the number of blue shirts is 54. By subtracting the number of green shirts (12) and the number of red shirts (28) from the total number of shirts (54), we can calculate the number of blue shirts, which make up the remaining portion of the total.

6. D: $\frac{31}{36}$

Set up the problem and find a common denominator for both fractions.

$$\frac{5}{12}+\frac{4}{9}$$

Multiply each fraction across by 1 to convert to a common denominator.

$$\frac{5}{12}\times\frac{3}{3}+\frac{4}{9}\times\frac{4}{4}$$

Once over the same denominator, add across the top. The total is over the common denominator.

$$\frac{15\ +16}{36}=\frac{31}{36}$$

7. D: Let *a* be the number of apples and *o* the number of oranges. Then, the total cost is $2a+3o=22$, while it also known that $a+o=10$. Using the knowledge of systems of equations, cancel the *o* variables by multiplying the second equation by -3. This makes the equation $-3a-3o=-30$. Adding this to the first equation, the b values cancel to get $-a=-8$, which simplifies to *a* = 8.

8. B: 8,685

Set up the problem, with the larger number on top. Begin subtracting with the far right column (ones). Borrow 10 from the column to the left, when necessary.

9. C: Each number in the sequence is adding one more than the difference between the previous two. For example, $10-6=4, 4+1=5$. Therefore, the next number after 10 is $10+5=15$. Going forward, $21-15=6, 6+1=7$. The next number is $21+7=28$. Therefore, the difference between numbers is the set of whole numbers starting at 2: 2, 3, 4, 5, 6, 7, ...

10. B: 37.797

Set up the problem, larger number on top and numbers lined up at the decimal. Begin subtracting with the far right column. Borrow 10 from the column to the left, when necessary.

1 **11. D:** This problem can be solved by using unit conversion. The initial units are miles per minute. The
2 final units need to be feet per second. Converting miles to feet uses the equivalence statement 1 mile
3 equals 5,280 feet. Converting minutes to seconds uses the equivalence statement 1 minute equals 60
4 seconds. Setting up the ratios to convert the units is shown in the following equation:

5

$$\frac{72\text{ mi}}{90\text{ min}}\times\frac{1\text{ min}}{60\text{ s}}\times\frac{5280\text{ ft}}{1\text{ mi}}=70.4\frac{\text{ft}}{\text{s}}$$

The initial units cancel out, and the new units are left.

12. A: $\frac{19}{24}$

Set up the problem and find a common denominator for both fractions.

$$\frac{23}{24} - \frac{1}{6}$$

Multiply each fraction across by 1 to convert to a common denominator.

$$\frac{23}{24} \times \frac{1}{1} - \frac{1}{6} \times \frac{4}{4}$$

Once over the same denominator, subtract across the top.

$$\frac{23 - 4}{24} = \frac{19}{24}$$

13. D: The volume of a cube is the length of the side cubed, and 3 inches cubed is 27 in^3. Choice *A* is not the correct answer because there was no operation performed. Choice *B* is not the correct answer because that is 2×3 inches. Choice *C* is not the correct answer because that is 3×3 inches.

14. D: The expression is three times the sum of twice a number and 1, which is $3(2x + 1)$. Then, 6 is subtracted from this expression.

15. D: $\frac{93}{8}$

The original number was $11\frac{5}{8}$. Multiply the denominator by the whole number portion. Add the numerator and put the total over the original denominator.

$$\frac{(8 \times 11) + 5}{8} = \frac{93}{8}$$

16. B: 30% is $\frac{3}{10}$. The number itself must be $\frac{10}{3}$ of 6, or $\frac{10}{3} \times 6 = 10 \times 2 = 20$.

17. C: $4\frac{1}{3} + 3\frac{3}{4} = 4 + 3 + \frac{1}{3} + \frac{3}{4} = 7 + \frac{1}{3} + \frac{3}{4}$. Adding the fractions gives $\frac{1}{3} + \frac{3}{4} = \frac{4}{12} + \frac{9}{12} = \frac{13}{12} = 1 + \frac{1}{12}$. Thus, $7 + \frac{1}{3} + \frac{3}{4} = 7 + 1 + \frac{1}{12} = 8\frac{1}{12}$.

18. A: The area of the square is the side length times itself, so $4 \times 4 = 16$ square inches. The area of a triangle is half the base times the height, so $\frac{1}{2} \times 2 \times 8 = 8$ square inches. The total is $16 + 8 = 24$ square inches.

19. B: Multiples of 6 can be found by skip-counting 6, 12, 18, 24, 30, etc. Multiples of 8 can be found by skip-counting 8, 16, 24, 32, etc. The first multiple that is common among the numbers is 24. The number 30 is a multiple of 6 but not of 8. To determine whether the other numbers are multiples of 6 and 8, each number can be divided by 6 and 8 to see if there is a remainder. The number 180 is divisible by 6 but not by 8. The number 440 is divisible by 8 but not by 6.

20. B: The perimeter is found by adding the value for all sides of a figure together. The perimeter of this figure is $5 + 4 + 2 + 2 + 3 + 3 + 6 + 13 = 38$.

21. D: The standard form for four hundred fifty-six thousand one hundred eighty-three is found by using the words to place the numbers in the correct place value. Here that number is 456,183.

22. C: From Week 3 to Week 4, the pea plant growth for each student is as follows: Chris, $2.1 - 1.7 = 0.4$; David, $1.9 - 1.6 = 0.3$; Emily, $2 - 1.4 = 0.6$; and Lauren, $2.25 - 1.75 = 0.5$. Emily's pea plant showed the most growth from Week 3 to Week 4.

23. A: This problem can be solved by simple multiplication and addition. Since the sale date is over six years apart, 6 can be multiplied by 12 for the number of months in a year, and then the remaining 4 months can be added.

$$(6 \times 12) + 4 = ?$$

$$72 + 4 = 76$$

24. D: In order to solve this problem, the number of feet in a yard must be established. There are 3 feet in every yard. The equation to calculate the minimum number of yards is $79 \div 3 = 26\frac{1}{3}$.

If the material is sold only by whole yards, then Mo would need to round up to the next whole yard in order to cover the extra $\frac{1}{3}$ yard. Therefore, the answer is 27 yards. None of the other choices meets the minimum whole yard requirement.

25. D: By grouping the four numbers in the answer into factors of the two numbers of the question (6 and 12), it can be determined that (3 x 2) x (4 x 3) = 6 x 12. Alternatively, each of the answer choices could be prime factored or multiplied out and compared to the original value. 6×12 has a value of 72 and a prime factorization of $2^3 \times 3^2$. The answer choices respectively have values of 64, 84, 108, 72, and 144 and prime factorizations of 2^6, $2^2 \times 3 \times 7$, $2^2 \times 3^3$, and $2^3 \times 3^2$, so answer *D* is the correct choice.

26. B: The formula for the perimeter of a rectangle is P=2l+2w, where P is the perimeter, l is the length, and w is the width. The first step is to substitute all of the data into the formula:

$$36 = 2(12) + 2W$$

Simplify by multiplying 2 x 12:

$$36 = 24 + 2W$$

Simplifying this further by subtracting 24 on each side, which gives:

$$36-24 = 24-24+2W$$

$$12= 2W$$

Divide by 2:

$$6 = W$$

The width is 6 cm. Remember to test this answer by substituting this value into the original formula:

$$36 = 2(12) + 2(6)$$

27. D: The addition rule is necessary to determine the probability because a 6 can be rolled on either roll of the die. The rule used is:

$$P(A \text{ or } B) = P(A) + P(B) - P(A \text{ and } B)$$

The probability of a 6 being individually rolled is $\frac{1}{6}$ and the probability of a 6 being rolled twice is:

$$\frac{1}{6} \times \frac{1}{6} = \frac{1}{36}$$

Therefore, the probability that a 6 is rolled at least once is:

$$\frac{1}{6} + \frac{1}{6} - \frac{1}{36} = \frac{11}{36}$$

28. A: To solve for the value of b, both sides of the equation need to be equalized.

Start by cancelling out the lower value of -4 by adding 4 to both sides:

$$5b - 4 = 2b + 17$$

$$5b - 4 + 4 = 2b + 17 + 4$$

$$5b = 2b + 21$$

The variable *b* is the same on each side, so subtract the lower 2b from each side:

$$5b = 2b + 21$$

$$5b - 2b = 2b + 21 - 2b$$

$$3b = 21$$

Then divide both sides by 3 to get the value of *b*:

$$3b = 21$$

$$\frac{3b}{3} = \frac{21}{3}$$

$$b = 7$$

29. D: To find the average of a set of values, add the values together and then divide by the total number of values. In this case, include the unknown value of what Dwayne needs to score on his next test, in order to solve it:

$$\frac{78 + 92 + 83 + 97 + x}{5} = 90$$

Add the unknown value to the new average total, which is 5. Then multiply each side by 5 to simplify the equation, resulting in:

$$78 + 92 + 83 + 97 + x = 450$$

$$350 + x = 450$$

$$x = 100$$

Dwayne would need to get a perfect score of 100 in order to get an average of at least 90.

Test this answer by substituting back into the original formula:

$$\frac{78 + 92 + 83 + 97 + 100}{5} = 90$$

30. D: For an even number of total values, the *median* is calculated by finding the *mean* or average of the two middle values once all values have been arranged in ascending order from least to greatest. In this case, $(92\ + 83) \div 2$ would equal the median 87.5, answer *D*.

Practice Test #3

Verbal Reasoning

Synonyms

Each of the questions below has one word. The one word is followed by four words or phrases. Please select one answer whose meaning is closest to the word in capital letters.

1. WEARY:
   a. clothing
   b. happy
   c. tired
   d. whiny

2. VAST:
   a. expansive
   b. ocean
   c. rapid
   d. small

3. DEMONSTRATE:
   a. build
   b. complete
   c. show
   d. tell

4. ORCHARD:
   a. farm
   b. fruit
   c. grove
   d. peach

5. TEXTILE:
   a. fabric
   b. knit
   c. mural
   d. ornament

6. OFFSPRING:
   a. bounce
   b. child
   c. music
   d. parent

7. PERMIT:
   a. allow
   b. crab
   c. law
   d. parking

8. INSPIRE:
   a. exercise
   b. impale
   c. motivate
   d. patronize

9. WOMAN:
   a. lady
   b. man
   c. mother
   d. women

10. ROTATION:
   a. flip
   b. spin
   c. wheel
   d. year

11. CONSISTENT:
   a. contains
   b. steady
   c. sticky
   d. texture

12. PRINCIPLE:
   a. foundation
   b. leader
   c. president
   d. principal

13. PERIMETER:
   a. area
   b. outline
   c. side
   d. volume

14. SYMBOL:
   a. clang
   b. drum
   c. emblem
   d. music

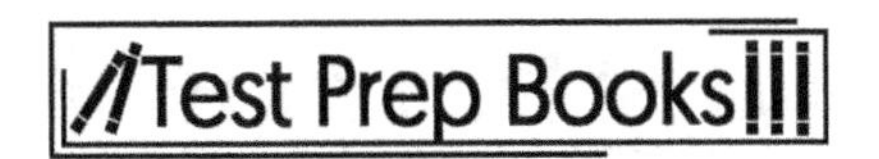

15. GERMINATE:
   a. doctor
   b. grow
   c. plants
   d. sick

16. OPPRESSED:
   a. acclaimed
   b. beloved
   c. helpless
   d. pressured

17. TRIUMPH:
   a. animosity
   b. banter
   c. burial
   d. celebration

Sentence Completion

Select the word or phrase that most correctly completes the sentence.

18. When the baseball game was over, the first thing Jackson did was run towards the dugout to grab his water bottle to relieve his __________ throat.
   a. dusty
   b. humid
   c. parched
   d. scorched

19. Driving across the United States, the two friends became __________ each time they arrived in a new state. They shared many good memories on that trip they would remember for the rest of their lives.
   a. closer
   b. distant
   c. irritable
   d. suffering

20. After Kira wrote her first book, she __________ her fans the sequel would be just as exciting as the first.
   a. denied
   b. germinated
   c. invigorated
   d. promised

21. When I heard the wolf howl from my tent, my hands started __________ and my heart stopped . . . hopefully I would make it through this night alive!
   a. dancing
   b. glowing
   c. shaking
   d. throbbing

22. Unlike Leo, who always played basketball in the park after school, Gabriel __________.
    a. ate his lunch in the cafeteria.
    b. would swim in the park after school.
    c. rode his bike to school in the morning.
    d. would usually go to the library and study after school.

23. As soon as the shot rang out, the runners __________ toward the finish line.
    a. herded
    b. rejoiced
    c. skipped
    d. sprinted

24. Determined to get an *A* on her paper, LaShonda __________.
    a. taught her little brother how to read.
    b. went to the gym everyday for a month.
    c. learned how to speak Spanish and French.
    d. began writing it two weeks before it was due.

25. After Colby's mom picked him up from school, they went to the bank to __________ a check.
    a. celebrate
    b. deposit
    c. eliminate
    d. neutralize

26. The sale at the grocery store __________ my dad to buy four avocados instead of two.
    a. berated
    b. dismayed
    c. inspired
    d. intimidated

27. My mom recently started drinking fruit and vegetable smoothies in order to __________.
    a. increase the quality of her health.
    b. obtain a raise at her new insurance job.
    c. prove to herself that she could hike the Appalachian trail.
    d. encourage her sister to start working out at the gym with her.

28. When Lindsay asked me to __________ her party, I immediately began writing a list of the birthday presents she might like to receive.
    a. acclaim
    b. amend
    c. astound
    d. attend

29. Cooking dinner was her favorite activity until she __________ the fire alarm by burning the casserole in the oven.
    a. activated
    b. disbanded
    c. offended
    d. unplugged

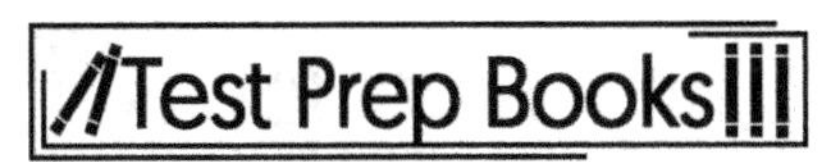

30. Before she arrived at the __________ dentist's office to take care of a cavity, she did some breathing exercises and made sure her teeth were clean.
    a. creative
    b. dreaded
    c. rapturous
    d. refreshing

31. The dog ran around in wide circles, disregarding the boy's command because he had not learned yet to be ____________.
    a. impudent
    b. obedient
    c. peaceful
    d. reserved

32. Ever since the bus changed its route from Anna's house to the other side of town, __________.
    a. Anna became afraid of the rain.
    b. Anna began riding her bike to school.
    c. Anna started receiving better grades.
    d. Anna proved to her friend she could beat her in Ping-Pong.

33. When we caught the eels, their bodies __________ out of our hands and back into the water.
    a. deteriorated
    b. exploded
    c. slithered
    d. thundered

34. Even though at the restaurant my mom __________ the eggplant with no cheese, she received a huge serving of parmesan on top.
    a. directed
    b. endorsed
    c. mourned
    d. requested

Quantitative Reasoning

1. In the following expression, which operation should be completed first? $5 \times 6 + 4 \div 2 - 1$.
    a. Addition
    b. Division
    c. Multiplication
    d. Subtraction

2. Which of the following is the definition of a prime number?
    a. A number less than 10
    b. A number divisible by 10
    c. A number that factors only into itself and one
    d. A number greater than one that factors only into itself and one

3. Which of the following is the correct order of operations?
 a. Exponents, Parentheses, Multiplication, Division, Addition, Subtraction
 b. Parentheses, Exponents, Addition, Multiplication, Division, Subtraction
 c. Parentheses, Exponents, Division, Addition, Subtraction, Multiplication
 d. Parentheses, Exponents, Multiplication, Division, Addition, Subtraction

4. If you were showing your friend how to round 245.2678 to the nearest thousandth, which place value would be used to decide whether to round up or round down?
 a. Ten-thousandth
 b. Thousandth
 c. Hundredth
 d. Thousand

5. Carey bought 184 pounds of fertilizer to use on her lawn. Each segment of her lawn required $12\frac{1}{2}$ pounds of fertilizer to do a sufficient job. If asked to determine how many segments could be fertilized with the amount purchased, what operation would be necessary to solve this problem?
 a. Addition
 b. Division
 c. Multiplication
 d. Subtraction

6. It is necessary to line up decimal places within the given numbers before performing which of the following operations?
 a. Division
 b. Fractions
 c. Multiplication
 d. Subtraction

7. Which of the following expressions best exemplifies the additive and subtractive identity?
 a. $9 - 9 = 0$
 b. $8 + 2 = 10$
 c. $6 + x = 6 - 6$
 d. $5 + 2 - 0 = 5 + 2 + 0$

8. Which four-sided shape is always a rectangle?
 a. Parallelogram
 b. Quadrilateral
 c. Rhombus
 d. Square

9. What unit of volume is used to describe the following 3-dimensional shape?

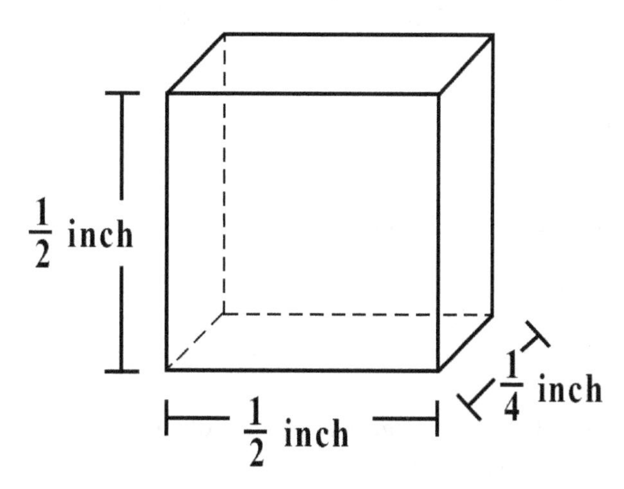

a. Cubic inches
b. Inches
c. Square inches
d. Squares

10. Which common denominator would be used in order to evaluate $\frac{2}{3}+\frac{4}{5}$?
a. 3
b. 5
c. 10
d. 15

11. In order to calculate the perimeter of a legal sized piece of paper that is 14 in and $8\frac{1}{2}$in wide, what formula would be used?
a. $P = 14 \times \frac{17}{2}$
b. $P = 14 + 8\frac{1}{2}$
c. $P = 14 \times 8\frac{1}{2}$
d. $P = 14 + 8\frac{1}{2} + 14 + 8\frac{1}{2}$

12. Which of the following are units in the metric system?
a. Inches, feet, miles, pounds
b. Kilograms, grams, kilometers, meters
c. Millimeters, centimeters, meters, pounds
d. Teaspoons, tablespoons, ounces

13. Which important mathematical property is shown in the following expression?

$$(7 \times 3) \times 2 = 7 \times (3 \times 2)$$

a. Associative property
b. Commutative property
c. Distributive property
d. Multiplicative inverse

14. The diameter of a circle measures 5.75 centimeters. What tool could be used to draw such a circle?
a. Compass
b. Meter stick
c. Ruler
d. Yard stick

15. Which of the following would be an instance in which ordinal numbers are used?
a. Katie scored a 9 out of 10 on her quiz.
b. Jacob missed one day of school last month.
c. Matthew finished second in the spelling bee.
d. Kim was 5 minutes late to school this morning.

16. Which of the following is represented by $6{,}000 + 400 + 30 + 6 + .5 + .08$?
a. 6,346.58
b. 6,436.058
c. 6,436.58
d. 64,365.8

17. What method is used to convert a fraction to a decimal?
a. Divide by 100 and reduce the fraction.
b. Multiply by 100 and reduce the fraction.
c. Divide the numerator by the denominator.
d. Divide the denominator by the numerator.

18. In Jim's school, there are 3 girls for every 2 boys. There are 650 students in total. Using this information, how many students are girls?
a. 65
b. 130
c. 260
d. 390

19. Emma works at a department store. She makes $50 per shift plus a commission of $5 on each sale (s) she makes during that shift. Which equation could be used to find out the total (t) Emma makes per shift?
a. $t = 5 + 50s$
b. $t = 50 + 5$
c. $t = 50 + 5s$
d. $t = 5 + 50 + s$

20. When evaluating word problems, which of the following phrases represent the division symbol?
    a. More than
    b. Product of
    c. Quotient of
    d. Results in

21. Which of the following choices results from solving the linear equation below?
$2(x - 9) = 2(3 - 2x)$
    a. $x = 3$
    b. $x = 3.75$
    c. $x = 4$
    d. $x = 8$

22. What would be the next term in the following sequence?
3, 9, 27, 81 . . .
    a. 90
    b. 135
    c. 243
    d. 323

23. Which two measurements of a triangle are needed to calculate the area of the triangle?
    a. Base, height
    b. Base, width
    c. Length, width
    d. Perimeter, height

24. Which of the following is a true statement regarding a line?
    a. A line has thickness.
    b. A line ends at two points.
    c. A line connects two points.
    d. A line and a line segment are the same thing.

25. What is an angle measuring less than 90 degrees called?
    a. Acute
    b. Complementary
    c. Obtuse
    d. Right

26. Which of the following units would be most appropriate to measure the size of a book?
    a. Feet
    b. Inches
    c. Millimeters
    d. Yards

27. Which of the following lists U.S. customary units for volume of liquids from largest to smallest?
    a. Fluid ounces, cup, pint, quart, gallon
    b. Gallon, quart, pint, cup, fluid ounces
    c. Gallon, quart, pint, fluid ounces, cup
    d. Quart, gallon, pint, cup, fluid ounces

28. Morgan wants to exercise at least 3 hours total this week. He exercised 35 minutes on Monday, 22 minutes on Tuesday, 41 minutes on Wednesday, and 1 hour on Thursday. How many more minutes does Morgan need to exercise to reach his goal of 3 hours?

a. 22
b. 25
c. 44
d. 81

29. Which of the following measure of central tendency is the data point in the middle of the sample when arranged in numerical order?

a. Mean
b. Median
c. Mode
d. Range

30. What is the mode of the following data set?

22, 18, 46, 37, 46, 25, 18, 33, 46, 25, 41

a. 18
b. 25
c. 33
d. 46

31. What is a data point with a value either extremely large or extremely small compared to the other values in a set?

a. Gap
b. Kurtosis
c. Outlier
d. Spread

32. In the following scatter plot, what type of correlation is present?

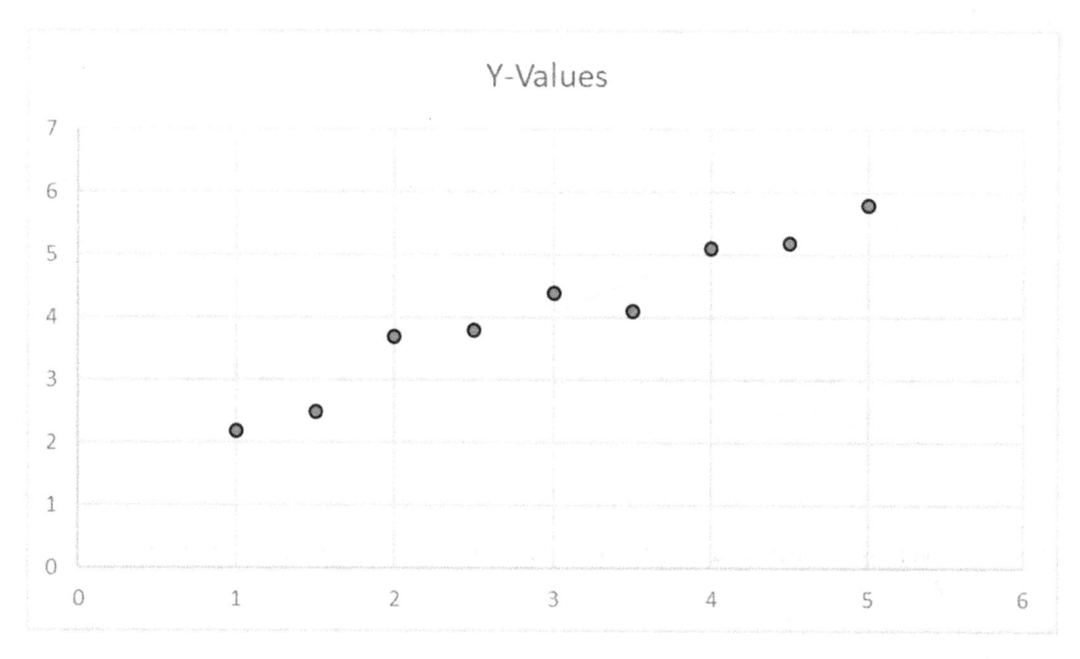

a. Independent
b. Negative
c. Positive
d. No correlation is present

33. In the following bar graph, how many children prefer brownies and cake?

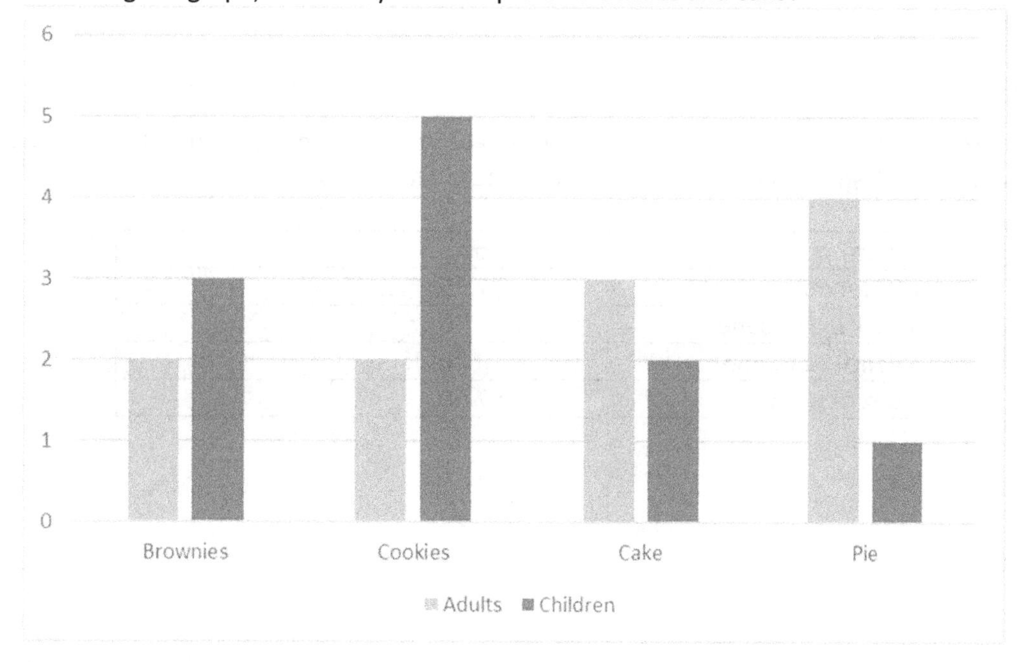

a. 3
b. 4
c. 5
d. 7

34. Which of the following graphs represents each category as a percentage of the whole data set?
    a. Bar graph
    b. Box plot
    c. Circle graph
    d. Line graph

35. Riley has 4 red shirts, 5 pink shirts, 2 blue shirts, 4 white shirts, and 1 black shirt. She pulls one out of her closet without looking. What is the probability she will pull out a white shirt?
    a. $\frac{1}{16}$
    b. $\frac{1}{8}$
    c. $\frac{3}{16}$
    d. $\frac{1}{4}$

36. What type of event is represented by a scenario where the first event does not affect the result of the second event?
    a. Dependent
    b. Independent
    c. Likely
    d. Unlikely

37. Which of the following is a three-dimensional shape?
    a. Circle
    b. Cube
    c. Hexagon
    d. Triangle

38. Trevor went to buy school supplies. He spent a total of $18.25. He bought 3 notebooks, 2 packages of paper, and 1 pencil sharpener. How many boxes of pens did he buy?

Item	Cost
Notebook	$2.75
Package of paper	$1.50
Pencil sharpener	$3.00
Box of pens	$2.00

    a. 2
    b. 3
    c. 4
    d. 5

Reading Comprehension

Questions 1–5 are based on the following passage.

1 Christopher Columbus is often credited for discovering America. This is incorrect. First, it is
2 impossible to "discover" something where people already live; however, Christopher Columbus
3 did explore places in the New World that were previously untouched by Europe, so the term
4 "explorer" would be more accurate. Another correction must be made, as well: Christopher
5 Columbus was not the first European explorer to reach the present day Americas! Rather, it was
6 Leif Erikson who first came to the New World and contacted the natives, nearly five hundred
7 years before Christopher Columbus.

8 Leif Erikson, the son of Erik the Red (a famous Viking outlaw and explorer in his own right), was
9 born in either 970 or 980, depending on which historian you seek. In 999, Leif left Greenland and
10 traveled to Norway where he would serve as a guard to King Olaf Tryggvason. It was there that
11 he became a convert to Christianity. Leif later tried to return home with the intention of taking
12 supplies and spreading Christianity to Greenland, however his ship was blown off course and he
13 arrived in a strange new land: present day Newfoundland, Canada".

14 When he finally returned to his adopted homeland Greenland, Leif consulted with a merchant
15 who had also seen the shores of this previously unknown land we now know as Canada. The son
16 of the legendary Viking explorer then gathered a crew of 35 men and set sail. Leif became the
17 first European to touch foot in the New World as he explored present-day Baffin Island and
18 Labrador, Canada. His crew called the land Vinland since it was plentiful with grapes. This
19 happened around 1000, nearly five hundred years before Columbus famously sailed the ocean
20 blue.

21 Eventually, in 1003, Leif set sail for home and arrived at Greenland with a ship full of timber. In
22 1020, seventeen years later, the legendary Viking died. Many believe that Leif Erikson should
23 receive more credit for his contributions in exploring the New World.

1. Which of the following best describes how the author generally presents the information?
   a. Cause-effect
   b. Chronological order
   c. Comparison-contrast
   d. Conclusion-premises

2. Which of the following is an opinion, rather than historical fact, expressed by the author?
   a. Leif Erikson's crew called the land Vinland since it was plentiful with grapes.
   b. Leif Erikson deserves more credit for his contributions in exploring the New World.
   c. Leif Erikson explored the Americas nearly five hundred years before Christopher Columbus.
   d. Leif Erikson was definitely the son of Erik the Red; however, historians debate the year of his birth.

3. Which of the following most accurately describes the author's main conclusion?
    a. Leif Erikson is a legendary Viking explorer.
    b. Leif Erikson contacted the natives nearly five hundred years before Columbus.
    c. Spreading Christianity motivated Leif Erikson's expeditions more than any other factor.
    d. Leif Erikson deserves more credit for exploring America hundreds of years before Columbus.

4. Which of the following best describes the author's intent in the passage?
    a. To alert
    b. To entertain
    c. To inform
    d. To suggest

5. Which of the following can be logically inferred from the passage?
    a. The Vikings disliked exploring the New World.
    b. Leif Erikson never shared his stories of exploration with the King of Norway.
    c. Historians have difficulty definitively pinpointing events in the Vikings' history.
    d. Leif Erikson's banishment from Iceland led to his exploration of present-day Canada.

Questions 6–10 are based on the following passage.

1 Smoking tobacco products is terribly destructive. A single cigarette contains over 4,000
2 chemicals, including 43 known carcinogens and 400 deadly toxins. Smoking can cause numerous
3 types of cancer including throat, mouth, nasal cavity, esophageal, gastric, pancreatic, renal,
4 bladder, and cervical cancer.

5 Cigarettes contain a drug called nicotine, one of the most addictive substances known to man.
6 Addiction is defined as a compulsion to seek the substance despite negative consequences.
7 According to the National Institute of Drug Abuse, nearly 35 million smokers expressed a desire
8 to quit smoking in 2015; however, more than 85 percent of those who struggle with addiction
9 will not achieve their goal. Almost all smokers regret picking up that first cigarette. You would be
10 wise to learn from their mistake if you have not yet started smoking.

11 According to the U.S. Department of Health and Human Services, 16 million people in the United
12 States presently suffer from a smoking-related condition and nearly nine million suffer from a
13 serious smoking-related illness. According to the Centers for Disease Control and Prevention
14 (CDC), tobacco products cause nearly six million deaths per year. This number is projected to
15 rise to over eight million deaths by 2030. Smokers, on average, die ten years earlier than their
16 nonsmoking peers.

17 In the United States, local, state, and federal governments typically tax tobacco products, which
18 leads to high prices. Nicotine users who struggle with addiction sometimes pay more for a pack
19 of cigarettes than for a few gallons of gas. Additionally, smokers tend to stink. The smell of
20 smoke is all-consuming and creates a pervasive nastiness. Smokers also risk staining their teeth
21 and fingers with yellow residue from the tar.

22 Smoking is deadly, expensive, and socially unappealing. Clearly, smoking is not worth the risks.

6. Which of the following statements most accurately summarizes the passage?
   a. Tobacco is less healthy than many alternatives.
   b. Tobacco products shorten smokers' lives by ten years and kill more than six million people per year.
   c. In the United States, local, state, and federal governments typically tax tobacco products, which leads to high prices.
   d. Tobacco is deadly, expensive, and socially unappealing, and smokers would be much better off kicking the addiction.

7. The author would be most likely to agree with which of the following statements?
   a. Other substances are more addictive than tobacco.
   b. Smokers should quit for whatever reason that gets them to stop smoking.
   c. Smokers should only quit cold turkey and avoid all nicotine cessation devices.
   d. People who want to continue smoking should advocate for a reduction in tobacco product taxes.

8. Which of the following represents an opinion statement on the part of the author?
   a. They also risk staining their teeth and fingers with yellow residue from the tar.
   b. Additionally, smokers tend to stink. The smell of smoke is all-consuming and creates a pervasive nastiness.
   c. Nicotine users who struggle with addiction sometimes pay more for a pack of cigarettes than a few gallons of gas.
   d. According to the Centers for Disease Control and Prevention (CDC), tobacco products cause nearly six million deaths per year.

9. What is the tone of this passage?
   a. Admiring
   b. Cautionary
   c. Indifferent
   d. Objective

10. What does the word *pervasive* mean in paragraph 4?
   a. A floral scent
   b. All over the place
   c. Pleasantly appealing
   d. To convince someone

This article discusses the famous poet and playwright William Shakespeare. Read it and answer questions 11–15.

1 People who argue that William Shakespeare is not responsible for the plays attributed to his
2 name are known as anti-Stratfordians (from the name of Shakespeare's birthplace, Stratford-
3 upon-Avon). The most common anti-Stratfordian claim is that William Shakespeare simply was
4 not educated enough or from a high enough social class to have written plays overflowing with
5 references to such a wide range of subjects like history, the classics, religion, and international
6 culture. William Shakespeare was the son of a glove-maker, he only had a basic grade school
7 education, and he never set foot outside of England—so how could he have produced plays of
8 such sophistication and imagination? How could he have written in such detail about historical
9 figures and events, or about different cultures and locations around Europe? According to anti-
10 Stratfordians, the depth of knowledge contained in Shakespeare's plays suggests a well-traveled
11 writer from a wealthy background with a university education, not a countryside writer like
12 Shakespeare. But in fact, there is not much substance to such speculation, and most anti-
13 Stratfordian arguments can be refuted with a little background about Shakespeare's time and
14 upbringing.

15 First of all, those who doubt Shakespeare's authorship often point to his common birth and brief
16 education as stumbling blocks to his writerly genius. Although it is true that Shakespeare did not
17 come from a noble class, his father was a very *successful* glove-maker and his mother was from
18 a very wealthy land owning family—so while Shakespeare may have had a country upbringing,
19 he was certainly from a well-off family and would have been educated accordingly. Also, even
20 though he did not attend university, grade school education in Shakespeare's time was actually
21 quite rigorous and exposed students to classic drama through writers like Seneca and Ovid. It is
22 not unreasonable to believe that Shakespeare received a very solid foundation in poetry and
23 literature from his early schooling.

24 Next, anti-Stratfordians tend to question how Shakespeare could write so extensively about
25 countries and cultures he had never visited before (for instance, several of his most famous
26 works like *Romeo and Juliet* and *The Merchant of Venice* were set in Italy, on the opposite side
27 of Europe!). But again, this criticism does not hold up under scrutiny. For one thing, Shakespeare
28 was living in London, a bustling metropolis of international trade, the most populous city in
29 England, and a political and cultural hub of Europe. In the daily crowds of people, Shakespeare
30 would certainly have been able to meet travelers from other countries and hear firsthand
31 accounts of life in their home country. And, in addition to the influx of information from world
32 travelers, this was also the age of the printing press, a jump in technology that made it possible
33 to print and circulate books much more easily than in the past. This also allowed for a freer flow
34 of information across different countries, allowing people to read about life and ideas from
35 throughout Europe. One needn't travel the continent in order to learn and write about its
36 culture.

11. Which sentence contains the author's thesis?
   a. People who argue that William Shakespeare is not responsible for the plays attributed to his name are known as anti-Stratfordians.
   b. It is not unreasonable to believe that Shakespeare received a very solid foundation in poetry and literature from his early schooling.
   c. Next, anti-Stratfordians tend to question how Shakespeare could write so extensively about countries and cultures he had never visited before.
   d. But in fact, there is not much substance to such speculation, and most anti-Stratfordian arguments can be refuted with a little background about Shakespeare's time and upbringing.

12. In the first paragraph, "How could he have written in such detail about historical figures and events, or about different cultures and locations around Europe?" is an example of which of the following?
   a. Hyperbole
   b. Onomatopoeia
   c. Rhetorical question
   d. Appeal to authority

13. How does the author respond to the claim that Shakespeare was not well-educated because he did not attend university?
   a. By insisting upon Shakespeare's natural genius.
   b. By explaining grade school curriculum in Shakespeare's time.
   c. By comparing Shakespeare with other uneducated writers of his time.
   d. By pointing out that Shakespeare's wealthy parents probably paid for private tutors.

14. The word "bustling" in the third paragraph most nearly means which of the following?
   a. Busy
   b. Expensive
   c. Foreign
   d. Undeveloped

15. According to the passage, what did Shakespeare's father do to make a living?
   a. He acted in plays.
   b. He was a cobbler.
   c. He was a king's guard.
   d. He was a glove-maker.

Questions 16–20 are based on the following passage.

1 The Myth of Head Heat Loss

2 It has recently been brought to my attention that most people believe that 75% of your body
3 heat is lost through your head. I had certainly heard this before, and am not going to attempt to
4 say I didn't believe it when I first heard it. It is natural to be gullible to anything said with enough
5 authority. But the "fact" that the majority of your body heat is lost through your head is a lie.

6 Let me explain. Heat loss is proportional to surface area exposed. An elephant loses a great deal
7 more heat than an anteater because it has a much greater surface area than an anteater. Each
8 cell has mitochondria that produce energy in the form of heat, and it takes a lot more energy to
9 run an elephant than an anteater.

10 So, each part of your body loses its proportional amount of heat in accordance with its surface
11 area. The human torso probably loses the most heat, though the legs lose a significant amount
12 as well. Some people have asked, "Why does it feel so much warmer when you cover your head
13 than when you don't?" Well, that's because your head, because it is not clothed, is losing a lot of
14 heat while the clothing on the rest of your body provides insulation. If you went outside with a
15 hat and pants but no shirt, not only would you look silly, but your heat loss would be
16 significantly greater because so much more of you would be exposed. So, if given the choice to
17 cover your chest or your head in the cold, choose the chest. It could save your life.

16. Why does the author compare elephants and anteaters?
    a. To express an opinion.
    b. To show the differences between them.
    c. To persuade why one is better than the other.
    d. To give an example that helps clarify the main point.

17. Which of the following best describes the tone of the passage?
    a. Angry
    b. Casual
    c. Harsh
    d. Indifferent

18. The author appeals to which branch of rhetoric to prove their case?
    a. Emotion
    b. Factual evidence
    c. Ethics and morals
    d. Author qualification

19. What does the word *gullible* mean?
    a. To be angry toward
    b. To be happy toward
    c. To distrust something
    d. To believe something easily

20. What is the main idea of the passage?
  a. To prove that you have to have a hat to survive in the cold.
  b. To debunk the myth that heat loss comes mostly from the head.
  c. To persuade the audience that anteaters are better than elephants.
  d. To convince the audience that heat loss comes mostly from the head.

Questions 21–25 are based on the following passage:

1 Children's literature holds a special place in many people's hearts. The stories that delight young
2 readers can be imaginative, educational, and help foster a love of reading in children. Stories
3 speak to children in a way adults sometimes do not understand. However, everything is not all
4 joy and happy endings in this genre. Many of the stories, tales, and books that are widely
5 recognized within this category feature darker themes. Some of the stories typically associated
6 with children, including fairy tales and fables, contain more serious issues such as child
7 abandonment, violence, and death. Some of the earliest fairytales come from *The Brothers*
8 *Grimm Fairytales.* These stories in their original form are surprisingly gruesome. Stories such as
9 "Hansel and Gretel," "Little Red Riding Hood," and "Cinderella" all contain elements that many
10 people would consider too dark for children. These early stories often presented these
11 disturbing images and elements in order to serve as a warning for children to induce good
12 behavior.

13 More recent entries into children's literature, such as *Where the Wild Things Are* and *The Giving*
14 *Tree*, are less shocking than some of the older tales but still touch on serious issues. When
15 children read about characters in a story dealing with these types of issues, it can help them
16 learn how to process some of the same emotions that occur in their own life. Whether children
17 are learning a lesson, processing emotions, or just enjoying a good story, reading literature can
18 be an important part of their life's journey.

21. What is the primary topic of this passage?
  a. Children's literature is universally loved.
  b. Children's literature cannot be enjoyed by adults.
  c. Children's literature often contains serious themes.
  d. Children's literature should only have happy endings.

22. What type of passage is this?
  a. Compare and contrast
  b. Descriptive
  c. Informative
  d. Persuasive

23. What is the meaning of the word "gruesome" in the first paragraph?
  a. Comfortable
  b. Horrible
  c. Numb
  d. Peculiar

24. What does the author mean to do by adding the following statement?

"These stories in their original form are surprisingly gruesome."

a. The stories were originally written in another language.
b. The original stories have been translated into different formats.
c. The intended audience for the early stories were adults instead of children.
d. When the stories were first written, they were grimmer than later adaptations.

25. What is meant by the figurative language in the following statement?
"Stories speak to children in a way adults sometimes do not understand."
a. Adults do not care for children's stories.
b. Children derive meaning from stories that adults do not.
c. Children's stories should always be read aloud to better understand the story.
d. Children's lack of life experiences causes them to sometimes misinterpret stories.

Math Achievement

1. Which of the following is equivalent to the value of the digit 3 in the number 792.134?

a. $\frac{3}{100}$

b. $\frac{3}{10}$

c. 3×10

d. 3×100

2. How will the following number be written in standard form:

$$(1 \times 10^4) + (3 \times 10^3) + (7 \times 10^1) + (8 \times 10^0)$$

a. 137
b. 1,378
c. 8,731
d. 13,078

3. How will the number 847.89632 be written if rounded to the nearest hundredth?
a. 847.89
b. 847.896
c. 847.90
d. 900

4. What is the value of the sum of $\frac{1}{3}$ and $\frac{2}{5}$?

a. $\frac{11}{30}$

b. $\frac{3}{8}$

c. $\frac{11}{15}$

d. $\frac{4}{5}$

5. What is the value of the expression: $7^2 - 3 \times (4 + 2) + 15 \div 5$?
   a. 12.2
   b. 34
   c. 40.2
   d. 58.2

6. How will $\frac{4}{5}$ be written as a percent?
   a. 40%
   b. 80%
   c. 90%
   d. 125%

7. If Danny takes 48 minutes to walk 3 miles, how long should it take him to walk 5 miles maintaining the same speed?
   a. 32 min
   b. 64 min
   c. 80 min
   d. 96 min

8. What are all the factors of 12?
   a. 12, 24, 36
   b. 1, 2, 4, 6, 12
   c. 12, 24, 36, 48
   d. 1, 2, 3, 4, 6, 12

9. A construction company is building a new housing development with the property of each house measuring 30 feet wide. If the length of the street is zoned off at 345 feet, how many houses can be built on the street?
   a. 11
   b. 11.5
   c. 12
   d. 115

10. What number could represent the point marked with a dot on the following number line?

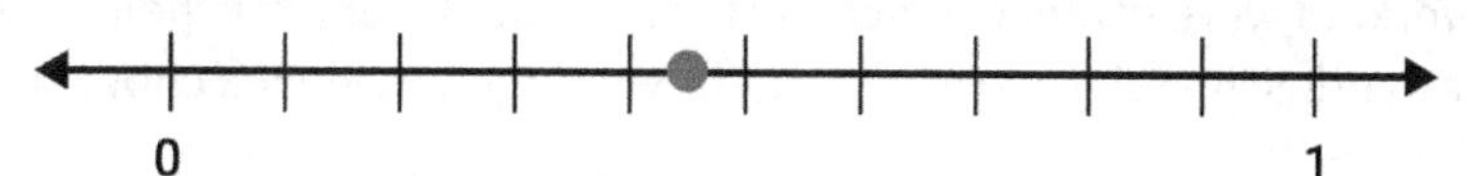

   a. $\frac{1}{4}$
   b. $\frac{2}{5}$
   c. .45
   d. .5

11. Kassidy drove for 3 hours at a speed of 60 miles per hour. Using the distance formula, $d = r \times t$ ($distance = rate \times time$), how far did Kassidy travel?
 a. 20 miles
 b. 65 miles
 c. 120 miles
 d. 180 miles

12. If $-3(x + 4) \geq x + 8$, what is the value of x?
 a. $x = 4$
 b. $x \geq 2$
 c. $x \geq -5$
 d. $x \leq -5$

13. Karen gets paid a weekly salary and a commission for every sale that she makes. The table below shows the number of sales and her pay for different weeks.

Sales	2	7	4	8
Pay	\$380	\$580	\$460	\$620

Which of the following equations represents Karen's weekly pay?
 a. $y = 90x + 200$
 b. $y = 90x - 200$
 c. $y = 40x + 300$
 d. $y = 40x - 300$

14. Which inequality represents the values displayed on the number line?

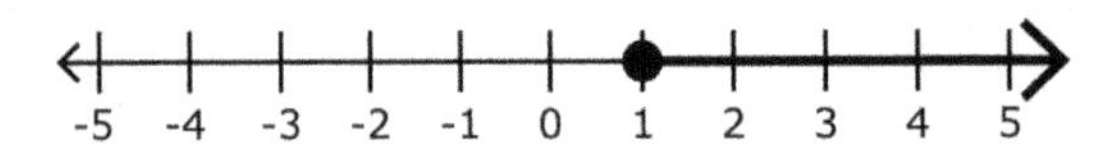

 a. $x < 1$
 b. $x \leq 1$
 c. $x > 1$
 d. $x \geq 1$

15. A group of 25 coworkers were given a choice of three lunch options for their upcoming meeting. If 12 employees chose a sandwich and 5 chose a salad, how many employees chose a hamburger?
 a. 5
 b. 8
 c. 13
 d. 17

16. Cindy makes four trips to the grocery store in March. She spends \$42.36, \$26.50, \$31.71, and \$37.23. What is the average amount Cindy spent on groceries in March?
 a. \$34
 b. \$34.45
 c. \$34.54
 d. \$35

17. Which of the following figures is not a polygon?
   a. Decagon
   b. Cone
   c. Rhombus
   d. Triangle

18. Joan is driving to her aunt's house for Thanksgiving. Her aunt lives 2.5 hours away. Joan stops at a rest stop for 15 minutes along the way. If Joan left at 10:15 a.m., what time will she arrive at her aunt's?
   a. 12 p.m.
   b. 12:45 p.m.
   c. 1:00 p.m.
   d. 1:15 p.m.

19. The area of a given rectangle is 24 square centimeters. If the measure of each side is multiplied by 3, what is the area of the new figure?
   a. 48 cm^2
   b. 72 cm^2
   c. 216 cm^2
   d. 13,824 cm^2

20. What are the coordinates of the point plotted on the grid?

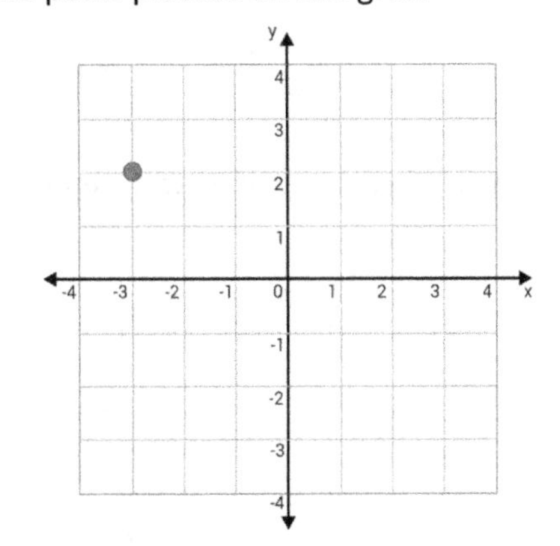

   a. (-3, 2)
   b. (2, -3)
   c. (-3, -2)
   d. (2, 3)

21. The perimeter of a 6-sided polygon is 56 cm. The length of three sides are 9 cm each. The length of two other sides are 8 cm each. What is the length of the missing side?
   a. 10 cm
   b. 11 cm
   c. 12 cm
   d. 13 cm

22. Katie works at a clothing company and sold 192 shirts over the weekend. One third of the shirts that were sold were patterned, and the rest were solid. Which mathematical expression would calculate the number of solid shirts Katie sold over the weekend?

a. $192 \times \frac{1}{3}$

b. $192 \div \frac{1}{3}$

c. $192 \div 3$

d. $192 \times (1 - \frac{1}{3})$

23. Which measure for the center of a small sample set is most affected by outliers?
a. Mean
b. Median
c. Mode
d. None of the above

24. Given the value of a given stock at monthly intervals, which graph should be used to best represent the trend of the stock?
a. Box plot
b. Circle graph
c. Line graph
d. Line plot

25. What is the probability of randomly picking the winner and runner-up from a race of 4 horses and distinguishing which is the winner?

a. $\frac{1}{16}$

b. $\frac{1}{12}$

c. $\frac{1}{4}$

d. $\frac{1}{2}$

26. Last year, the New York City area received approximately $27\frac{3}{4}$ inches of snow. The Denver area received approximately 3 times as much snow as New York City. How much snow fell in Denver?

a. $9\frac{1}{4}$ inches

b. $27\frac{1}{4}$ inches

c. 60 inches

d. $83\frac{1}{4}$ inches

27. Evaluate $9 \times 9 \div 9 + 9 - 9 \div 9$.
    a. 0
    b. 9
    c. 17
    d. 81

28. Alan currently weighs 200 pounds, but he wants to lose weight to get down to 175 pounds. What is this difference in kilograms? (1 pound is approximately equal to 0.45 kilograms.)
    a. 9 kg
    b. 11.25 kg
    c. 78.75 kg
    d. 90 kg

29. Johnny earns $2334.50 from his job each month. He pays $1437 for monthly expenses. Johnny is planning a vacation in 3 months that he estimates will cost $1750 total. How much will Johnny have left over from three months of saving once he pays for his vacation?
    a. $584.50
    b. $852.50
    c. $942.50
    d. $948.50

30. Solve the following:

$$4 \times 7 + (25 - 21)^2 \div 2$$

    a. 22
    b. 36
    c. 60.5
    d. 512

Essay

Select a topic from the list below and write an essay. You may organize your essay on another sheet of paper.

Topic 1: Who is someone you look up to the most in this world? Why?

Topic 2: If you could move anywhere in the world, where would it be and why?

Topic 3: If you could have dinner with anyone in the world, alive or dead, who would it be? Why have you chosen this person?

Answer Explanations #3

Verbal Reasoning

Synonyms

1. C: Weary most closely means tired. Someone who is weary and tired may be whiny, but they do not necessarily mean the same thing.

2. A: Something that is vast is big and expansive. Choice *B*, ocean, may be described as vast. However, the word itself does not mean vast. The heavens or skies may also be described as vast. Someone's imagination or vocabulary can also be vast.

3. C: To demonstrate something means to show it. A demonstration is a show-and-tell type of example. It is usually visual.

4. C: An orchard is most like a grove. Both are areas like plantations that grow different kinds of fruit. Peach is a type of fruit that may be grown in an orchard. However, *peach* is not a synonym for orchard. Many citrus fruits are grown in groves. But either word can be used to describe many fruit-bearing trees in one area. Choice *A*, farm, may have an orchard or grove on the property. However, they are not the same thing, and many farms do not grow fruit trees.

5. A: A textile is another word for a fabric. The most confusing choice in this case is Choice *B*, knit. This is because some textiles are knit, but *textile* and *knit* are not synonyms. Plenty of textiles are not knit.

6. B: Offspring are the kids of parents. This word is common when talking about the animal kingdom, though it can be used with humans as well. *Offspring* does have the word *spring* in it. But it has nothing to do with bounce, Choice *A*. Choice *D*, parent, maybe tricky because parents have offspring. But for this reason, they are not synonyms.

7. A: Permit can be a verb or a noun. As a verb, it means to allow or give permission for something. As a noun, it refers to a document or something that has been authorized like a parking permit or driving permit. This would allow the authorized person to park or drive under the rules of the document.

8. C: If someone is inspired, they are driven to do something. Someone who is an inspiration motivates others to follow their lead.

9. A: A woman is a lady. You must read carefully and remember the difference between *woman* and *women*. *Woman* refers to one person who is female. W*omen* is the plural form and refers to more than one, or a group, of ladies. A woman can be a mother, but not necessarily. *Woman* and *mother* are not synonyms.

10. B: Rotation means to spin or turn, like a wheel rotating on a car. But *wheel*, Choice *C*, does not mean the same thing as the word *rotation*.

11. B: Something that is consistent is steady, predictable, reliable, or constant. The tricky ones here is that the word *consistency* comes from the word consistent. *Consistency* may describe a texture or something that is sticky, Choices *C* and *D*. *Consistent* also comes from the word *consist. Consist* means to contain (Choice *A*).

12. A: A principle is a guiding idea or belief. Someone with good moral character is described as having strong principles. You must be careful not to get confused with the homonyms *principle* and *principal*, Choice *D*. These two words have different meanings. A principal is the leader of a school. The word principal also refers to the main idea or most important thing.

13. B: Perimeter refers to the outline of an object. You may recognize that word from math class. In math class, perimeter refers to the edges or distance around a closed shape. Some of the other choices refer to other math words encountered in geometry. However, they do not have the same meaning as *perimeter.*

14. C: A symbol is an object, picture, or sign that is used to represent something. For example, a pink ribbon is a symbol for breast-cancer awareness. A flag can be a symbol for a country. The tricky part of this question was also knowing the meaning of *emblem. Emblem* describes a design that represents a group or concept, much like a symbol. Emblems often appear on flags or a coat of arms.

15. B: Germinate means to develop or grow. It most often refers to sprouting seeds as a new plant first breaks through the seed coat. It can also refer to the development of an idea. Choice *C, plants*, may be an attractive choice since plants germinate. However, the word *germinate* does not mean *plant.*

16. C: The word *oppressed* means being exploited or helpless, Choice *C*. Choice *A*, acclaimed, means being praised. To be beloved, Choice *B*, means to be cherished and adored. To be pressured, Choice *D*, means to be pushed into doing something, in some contexts.

17. D: The word triumph most closely means celebration, Choice *D*. Animosity, Choice *A*, means strong dislike or hatred, and is very different from the word *triumph*. Choice *B*, banter, is the act of teasing. Burial, Choice *C*, is the act of burying the dead.

Sentence Completion

18. C: Jackson wanted to relieve his *parched* throat. *Parched* is the correct answer because it means *thirsty*. While Jackson's throat could have been dusty, Choice *A*, from playing baseball, one usually doesn't need to relieve a dusty throat, but instead clean it. Choice *B*, humid, means moist, and usually refers to the weather. Choice *D*, scorched, means blackened or baked, and doesn't fit in this context.

19. A: The two friends became closer. For this question, it's important to look at the context of the sentence. The second sentence says the friends shared good memories on the trip, which would not make the friends distant or irritable, Choices *B* and *C*. Choice *D* does not grammatically fit within the sentence: "became suffering" is incorrect usage. Therefore, Choice *A* is correct.

20. D: She promised her fans the sequel would be just as exciting as the first. Choice *A*, denied, is the opposite of the word *promised* and does not fit with the word *excited*. Choice *B*, germinated, means to grow. Choice *C*, invigorated, means energized, and might fit the tone of the sentence with the word *excited.* However, *promised* is the better word to use here.

21. C: My hands started shaking and my heart stopped. Usually when someone is afraid or nervous, their hands start to shake. Choice *A*, dancing, does not make sense in the context of the sentence. Choice *B*, glowing, is incorrect; hands usually do not glow when one is afraid of something. Choice *D*, throbbing, is closer than *A* or *B*, but Choice *C*, shaking, is a better answer than *throbbing*.

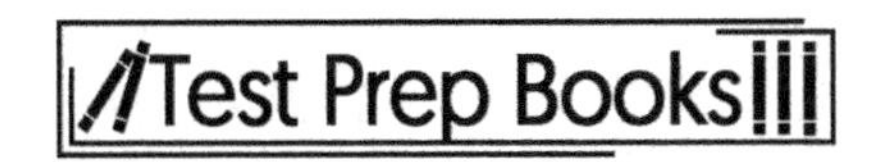

22. D: Gabriel would usually go to the library and study after school. The word *unlike* tells us that Gabriel would do the opposite of what Leo would do at the time. Choices *A* and *C* talk about different times, in the morning and at lunch, so they are not the best answer choice here. Choice *B* is more similar to what Leo would do than Choice *D*, so Choice *D* is the correct choice.

23. D: The runners sprinted toward the finish line. Choice *A*, herded, means to gather around something; usually *herded* is used for animals and not for runners. Choice *B* does not fit within the context of the sentence, as normally runners would be *sprinting* and not *rejoicing* toward a finish line. Choice *C* is incorrect; runners who begin a race usually don't skip toward the finish line.

24. D: LaShonda began writing it two weeks before it was due. Choices *A, B,* and *C* all reference activities that don't have anything to do with writing a paper. If her paper was about any one of these things, like going to the gym or teaching someone to read, then perhaps these would be decent answers. However, we are not given enough information to know what the topic of the paper is. Therefore, Choice *D* is the correct answer.

25. B: They went to the bank to deposit a check. When people go to the bank with a check, they usually don't celebrate it, Choice *A*, but do something more practical with it, like deposit it. Choice *C*, eliminate, means to get rid of something, and is also incorrect here. Choice *D*, neutralize, means to counteract something, and is incorrect in this sentence.

26. C: The sale at the grocery store inspired my dad to buy four avocados instead of two. The word *inspired* means to encourage or stimulate. Sales are usually seen as a positive experience, so the sale *inspired* the dad to buy more avocados. The rest of the words (berated, dismayed, and intimidated) have negative connotations and therefore do not fit within the context of the sentence.

27. A: In order to increase the quality of her health. This phrase makes the most sense within the context of the sentence. Becoming healthier is a direct effect of consuming fruits and vegetables, so Choice *A* is the best answer choice here. Becoming healthier might lead to Choices *B, C,* and *D*, but these are not the best answers for this question.

28. D: When Lindsay asked me to attend her party. Choice *A*, acclaim, is an expression of approval, and is not the right fit here. Choice *B*, amend, means to improve or correct. Again, this is not the best choice for the context; the speaker is wondering what gift to bring Lindsay, and is not thinking about ways to correct the party. Choice *C*, astound, means to amaze. Usually we don't hear of people "amazing" other people's parties. This is not the best choice.

29. A: Until she activated the fire alarm by burning the casserole in the oven. Choice *B*, disbanded, is the opposite of *activated* and is incorrect in this context. Choice *C*, offended, is the wrong choice here. You can offend a person because they have emotions, but you cannot offend a fire alarm. Choice *D*, unplugged, is also incorrect. You cannot unplug a fire alarm by burning a casserole.

30. B: Before she arrived at the dreaded dentist's office. Choices *A*, creative, does not fit within the context of the sentence. Choice *C*, rapturous, means ecstatic or happy, and is the opposite sentiment of what we are looking for. Choice *D*, refreshing, is not an adjective used to describe a dentist's office, especially when the patient is about to take care of a cavity.

31. B: *Obedient* means well-behaved. The dog hasn't learned to heed commands yet, so he is not obedient. Impudent means bold which does not fit with the rest of the sentence. Peaceful and reserved are related to good behavior, but obedient is a better choice to complete the sentence.

32. B: Anna began riding her bike to school. This question determines whether or not you can understand the nature of cause and effect. Choices *A*, *C*, and *D* could possibly be in a chain of events of effects from the bus taking a different route. However, the most direct cause is Choice *B*.

33. C: Their bodies slithered out of our hands and back into the water. Deteriorated, Choice *A*, means to crumble or disintegrate, which is not something an eel's body is likely to do on its own. Choices *B* and *D*, exploded and thundered, are too extreme for the context. Choice *C* is the best answer for this sentence.

34. D: My mom requested the eggplant with no cheese. Directed, Choice A, means to supervise or conduct, and does not make sense in the context of the sentence. Choice *B*, endorsed, means to approve or support something. Pay attention to words like *even though* that suggest a contrast of surprising facts. Choice *C*, mourned, means to grieve over, and is also incorrect.

Quantitative Reasoning

1. C: Using the order of operations, multiplication and division are computed first from left to right. Multiplication is on the left; therefore, multiplication should be performed first.

2. D: A number is prime because its only factors are itself and one. Positive numbers (greater than zero) can be prime numbers.

3. D: Order of operations follows PEMDAS—Parentheses, Exponents, Multiplication and Division from left to right, and Addition and Subtraction from left to right.

4. A: The place value to the right of the thousandth place, which would be the ten-thousandth place, is what gets utilized. The value in the thousandth place is 7. The number in the place value to its right is greater than 4, so the 7 gets bumped up to 8. Everything to its right turns to a zero, to get 245.2680. The zero is dropped because it is part of the decimal.

5. B: This is a division problem because the original amount needs to be split up into equal amounts. The mixed number $12\frac{1}{2}$ should be converted to an improper fraction first.

$$12\frac{1}{2} = (12 \times 2) + \frac{1}{2} = \frac{23}{2}$$

Carey needs to determine how many times $\frac{23}{2}$ goes into 184. This is a division problem:

$$184 \div \frac{23}{2} = ?$$

The fraction can be flipped, and the problem turns into the multiplication:

$$184 \times \frac{2}{23} = \frac{368}{23}$$

This improper fraction can be simplified into 16 because:

$$368 \div 23 = 16$$

The answer is 16 lawn segments.

6. D: Numbers should be lined up by decimal places before subtraction is performed. This is because subtraction is performed within each place value. The other operations, such as multiplication, division, and exponents (which is a form of multiplication), involve ignoring the decimal places at first and then including them at the end.

7. D: The additive and subtractive identity is zero. When added or subtracted to any number, zero does not change the original number.

8. D: A rectangle is a specific type of parallelogram. It has 4 right angles. A square is a rhombus that has 4 right angles. Therefore, a square is always a rectangle because it has two sets of parallel lines and 4 right angles.

9. A: Volume of this 3-dimensional figure is calculated using length x width x height. Each measure of length is in inches. Therefore, the answer would be labeled in cubic inches.

10. D: A common denominator must be found. The least common denominator is 15 because it has both 5 and 3 as factors. The fractions must be rewritten using 15 as the denominator.

11. D: Perimeter of a rectangle is the sum of all four sides. Therefore, the answer is:

$$P = 14 + 8\,{}^{1}/_{2} + 14 + 8\,{}^{1}/_{2}$$

$$14 + 14 + 8 + {}^{1}/_{2} + 8 + {}^{1}/_{2}$$

45 inches

12. B: Inches, pounds, and baking measurements, such as tablespoons, are not part of the metric system. Kilograms, grams, kilometers, and meters are part of the metric system.

13. A: It shows the associative property of multiplication. The order of multiplication does not matter, and the grouping symbols do not change the final result once the expression is evaluated.

14. A: A compass is a tool that can be used to draw a circle. The compass would be drawn by using the length of the radius, which is half of the diameter.

15. C: Ordinal numbers represent a ranking. Placing second in a competition is a ranking among the other participants of the spelling bee.

16. C: The expanded or decomposed form represents the sum of each place value of a number. The numbers are added back together to find the standard form of the number.

17. C: The method used to convert a fraction to a decimal is to divide the numerator by the denominator.

18. D: Three girls for every two boys can be expressed as a ratio: 3:2. This can be visualized as splitting the school into 5 groups: 3 girl groups and 2 boy groups. The number of students which are in each group can be found by dividing the total number of students by 5:

$$\frac{650 \text{ students}}{5 \text{ groups}} = \frac{130 \text{ students}}{\text{group}}$$

To find the total number of girls, multiply the number of students per group (130) by the number of girl groups in the school (3). This equals 390, Choice *D*.

19. C: Algebraic expressions can be used with unknown quantities to create equations that represent real-world situations. In this example, the $50 per shift amount would be added to the commission amount of $5 times the amount of sales per shift to reach the total.

20. C: Phrases such as divided by, quotient of, or half of can all be used to indicate the use of the division symbol in a mathematical expression. More than typically indicates the addition symbol. The multiplication symbol is represented by the phrase product of. Results in is another way to say equal to and is shown by the equal sign.

21. C: The linear equation can be solved by first using the distributive property which results in:

$$2x - 18 = 6 - 4x$$

The next step is to combine the like terms resulting in:

$$6x = 24$$

The final step is to divide both sides by 6 to find:

$$x = 4$$

22. C: The sequence presented is a geometric sequence where each step is multiplied by 3 to get the next subsequent step. $81 \times 3 = 243$, so the correct answer is 243.

23. A: The correct formula for calculating the area of a triangle is $A = \frac{1}{2} \times b \times h$, where A is equal to area, b is equal to base, and h is equal to height. Therefore, the two measurements needed to calculate the area of a triangle are base and height.

24. C: A line also no thickness. If the line does end at the two points, then it is a line segment. A line does connect two points, but it extends indefinitely in both directions.

25. A: An angle measuring less than 90 degrees is called an acute angle. A complementary angle is one of two angles which when added together equal 90 degrees. An angle measuring more than 90 degrees is an obtuse angle. A right angle is an angle that measures 90 degrees.

26. B: The most appropriate unit to use when measuring the size of a book is inches. This is based on the knowledge of the approximate size of most books and knowing the approximate size of the unit of measurement. Both yards and feet are too large, while millimeters are too small.

27. B: The U.S. customary units of measurement for volume of liquids is from largest to smallest: gallon, quart, pint, cup, and fluid ounces.

28. A: Morgan will need to do 22 more minutes of exercise to reach his goal of 3 hours for the week. The following equation can be used to find the solution.

$$\left(3\ hours \times \frac{60\ minutes}{1\ hour}\right) - 35\ minutes - 22\ minutes - 41\ minutes - \left(1\ hour \times \frac{60\ minutes}{1\ hour}\right)$$

$$22\ minutes$$

The information given in hours must be converted to minutes so the amounts can be used to solve the equation.

29. B: The median in a data set is the data point found in the middle of the set when arranged in numerical order from least to greatest. The mean is the average of a data set. The mode is the value that occurs the most in a data set. Range refers to the spread of values in a data set.

30. D: The mode of a data set is 46 because it appears most often. In this data set, there are three 46's which is more than any other value. 18 and 25 both appear only twice in the sample. 33 is the median of the data set.

31. C: A data point that is very large or very small comparatively to the other values in the data set is an outlier. A gap is an interval in the range of a data set where there are no data points. Kurtosis measures whether data has a high or low number of outliers. Spread represents how far data points are from the center of a set.

32. C: The correlation between the two variables in this scatter plot is positive because the line of best-fit that can be drawn would have a positive slope. Independent refers to the variables on the x-axis and does not address correlation. If the correlation was negative, the line of best-fit would have a negative slope. If the data points were not in a linear formation, and no line of best-fit could be drawn, there would be no correlation.

33. C: In this bar graph, 5 children prefer brownies and cake. The children columns for brownies and cake have heights of 3 and 2, respectively. Added together that makes a total of 5 children.

34. C: A circle graph does not show the frequency of the individual categories, but rather indicates how much of the whole set (the percentage) is represented by each category. A bar graph and a line graph both demonstrate the frequency of each category within a data set. A box plot divides a data set into quartiles.

35. D: There are 16 shirts in all and 4 white shirts, so the probability of randomly choosing a white shirt is:

$$\frac{4}{16} = \frac{1}{4}$$

36. B: Independent events are situations where the outcome of the first event has no impact on the outcome of the second event. Dependent events have outcomes that do affect one another. Likely and unlikely refer to the chance an event will occur or not.

37. B: A cube is a three-dimensional solid shape. Two-dimensional shapes are flat like circles, triangles, hexagons, squares, and other similar shapes.

38. A: Trevor bought 2 boxes of pens. If he bought 3 notebooks, 2 packages of paper, and 1 pencil sharpener, then he spent $8.25, $3, and $3 on those items respectively. The total of those items then is $14.25. The difference between the total of those items and the total spent of $18.25 is $4. Therefore, he bought 2 boxes of pens at $2 apiece for $4.

Reading Comprehension

1. D: The passage does not proceed in chronological order since it begins by pointing out Leif Erikson's explorations in America so Choice *B* does not work. Although the author compares and contrasts Erikson with Christopher Columbus, this is not the main way the information is presented; therefore, Choice *C* does not work. Neither does Choice *A* because there is no mention of or reference to cause and effect in the passage. However, the passage does offer a conclusion (Leif Erikson deserves more credit) and premises (first European to set foot in the New World and first to contact the natives) to substantiate Erikson's historical importance. Thus, Choice *D* is correct.

2. B: Choice *A* is wrong; that Erikson called the land Vinland is a verifiable fact as is Choice *C* because he did contact the natives almost 500 years before Columbus. Choice *D* is wrong because it describes facts: Leif Erikson was the son of Erik the Red and historians debate Leif's date of birth. These are not opinions. Choice *B* is the correct answer because it is the author's opinion that Erikson deserves more credit. That, in fact, is his conclusion in the piece, but another person could argue that Columbus or another explorer deserves more credit for opening up the New World to exploration. Rather than being an incontrovertible fact, it is a subjective value claim.

3. D: Choice *A* is wrong because the author aims to go beyond describing Erikson as a mere legendary Viking. Choice *B* is wrong because it is a premise that Erikson contacted the natives 500 years before Columbus, which is simply a part of supporting the author's conclusion. Choice *C* is wrong because the author does not focus on Erikson's motivations, let alone name the spreading of Christianity as his primary objective. Choice *D* is correct because, as stated in the previous answer, it accurately identifies the author's statement that Erikson deserves more credit than he has received for being the first European to explore the New World.

4. C: Although the author is certainly trying to alert the readers of Leif Erikson's unheralded accomplishments, the nature of the writing does not indicate the author would be satisfied with the reader merely knowing of Erikson's exploration (Choice *A*). Choice *B* is wrong because the author is not in any way trying to entertain the reader. Choice *D* is wrong because he goes beyond a mere suggestion; "suggest" is too vague. Rather, the author would want the reader to be informed about it, which is more substantial (Choice *C*).

5. C: Choice *A* is wrong because the author never addresses the Vikings' state of mind or emotions. Choice *B* is wrong because there is not enough information to support this premise. It is unclear whether Erikson informed the King of Norway of his finding. Although it is true that the King did not send a follow-up expedition, he could have simply chosen not to expend the resources after receiving Erikson's news. It is not possible to logically infer whether Erikson told him. Choice *D* is wrong because the author does not elaborate on Erikson's exile and whether he would have become an explorer if not for his banishment. Choice *C* is correct because there are two examples—Leif Erikson's date of birth and what happened during the encounter with the natives—of historians having trouble pinning down important dates in Viking history.

6. D: The author is opposed to tobacco. He cites disease and deaths associated with smoking. He points to the monetary expense and aesthetic costs. Choice *A* is incorrect because alternatives to smoking are not even addressed in the passage. Choice *B* is incorrect because, while these statistics are a premise in the argument, they do not represent a summary of the piece. Choice *C* is incorrect because it does not summarize the passage but rather is just a premise. Choice *D* is the correct answer because it states the three critiques offered against tobacco and expresses the author's conclusion.

7. C: We are looking for something the author would agree with, so it will almost certainly be anti-smoking or an argument in favor of quitting smoking. Choice *A* is incorrect because the author does not reference other substances but does speak of how addictive nicotine, a drug in tobacco, is. Choice *C* is incorrect because the author does not speak against means of cessation. Choice *D* is incorrect because the author certainly would not encourage reducing taxes to encourage a reduction of smoking costs, thereby helping smokers to continue the habit. Choice *B* is correct because the author is definitely attempting to persuade smokers to quit smoking.

8. B: Here, we are looking for an opinion of the author's rather than a fact or statistic. Choice *A* is incorrect because yellow stains are a known possible adverse effect of smoking. Choice *C* is incorrect because it expresses the fact that cigarettes sometimes cost more than a few gallons of gas. It would be an opinion if the author said that cigarettes were not affordable. Choice *D* is incorrect because quoting statistics from the Centers of Disease Control and Prevention is stating facts, not opinions. Choice *B* is correct as an opinion because smell is subjective. Some people might like the smell of smoke, they might not have working olfactory senses, and/or some people might not find the smell of smoke akin to "pervasive nastiness," so this is the expression of an opinion. Thus, Choice *B* is the correct answer.

9. B: The passage is cautionary, because the author warns about the hazards of smoking and uses the second-person "you" to offer suggestions, like "You would be wise to learn from their mistake." Choice *A* is incorrect; the passage is opposite of admiring towards the subject of smoking. Choice C, indifferent, is incorrect because the author expresses an opinion and makes it clear they dislike smoking. Choice *D*, objective, means that the passage would be totally without persuasion or suggestions, so this answer choice is incorrect.

10. B: The word *pervasive* means "all over the place." The passage says that "The smell of smoke is all-consuming and creates a *pervasive* nastiness," which means a smell that is *everywhere* or *all over*. Choices *A* and *C*, pleasantly appealing and a floral scent, are too pleasant for the context of the passage. Choice *D* doesn't make sense in the sentence, as "to convince someone" wouldn't really describe the word *nastiness* like pervasive does.

11. D: But in fact, there is not much substance to such speculation, and most anti-Stratfordian arguments can be refuted with a little background about Shakespeare's time and upbringing. The thesis is a statement that contains the author's topic and main idea. The main purpose of this article is to use historical evidence to provide counterarguments to anti-Stratfordians. Choice *A* is simply a definition; Choice *B* is a supporting detail, not a main idea; and Choice *C* represents an idea of anti-Stratfordians, not the author's opinion.

12. C: Rhetorical question. This requires readers to be familiar with different types of rhetorical devices. A rhetorical question is a question that is asked not to obtain an answer but to encourage readers to more deeply consider an issue.

13. B: By explaining grade school curriculum in Shakespeare's time. This question asks readers to refer to the organizational structure of the article and demonstrate understanding of how the author provides details to support their argument. This particular detail can be found in the second paragraph: "even though he did not attend university, grade school education in Shakespeare's time was actually quite rigorous."

14. A: Busy. This is a vocabulary question that can be answered using context clues. Other sentences in the paragraph describe London as "the most populous city in England" filled with "crowds of people," giving an image of a busy city full of people. Choice *B* is not mentioned in the passage. Choice *C* is incorrect because London was in Shakespeare's home country, not a foreign one. Choice *D* is not a good answer choice because the passage describes how London was a popular and important city, probably not an undeveloped one.

15. C: Shakespeare's father was a glove-maker. The passage states this fact in paragraph two, where it says "his father was a very successful glove-maker." The other answer choices are incorrect.

16. D: Choice *D* is correct because the author is trying to demonstrate the main idea, which is that heat loss is proportional to surface area, and so they compare two animals with different surface areas to clarify the main point. Choice *A* is incorrect because the author uses elephants and anteaters to prove a point, that heat loss is proportional to surface area, not to express an opinion. Choice *B* is incorrect because though the author does use them to show differences, they do so in order to give examples that prove the above points, so Choice *B* is not the best answer. Choice *C* is incorrect because there is no language to indicate favoritism between the two animals.

17. B: Because of the way that the author addresses the reader, and also the colloquial language that the author uses (i.e., "let me explain," "so," "well," didn't," "you would look silly," etc.), *B* is the best answer because it has a much more casual tone than the usual informative article. *A* is incorrect because again, while not necessarily nice, the language does not carry an angry charge. Choice *C* may be a tempting choice because the author says the "fact" that most of one's heat is lost through their head is a "lie," and that someone who does not wear a shirt in the cold looks silly, but it only happens twice within all the diction of the passage and it does not give an overall tone of harshness. The author is clearly not indifferent to the subject because of the passionate language that they use, so *D* is incorrect.

18. B: The author gives logical examples and reasons in order to prove that most of one's heat is not lost through their head, therefore *B* is correct. *A* is incorrect because there is not much emotionally charged language in this selection, and even the small amount present is greatly outnumbered by the facts and evidence. *C* is incorrect because there is no mention of ethics or morals in this selection. *D* is incorrect because the author never qualifies himself as someone who has the authority to be writing on this topic.

19. D: *Gullible* means to believe something easily. The other answer choices could fit easily within the context of the passage: you can be angry toward, distrustful toward, or happy toward authority. For this answer choice and the surrounding context, however, the author talks about a myth that people believe easily, so *gullible* would be the word that fits best in this context.

20. B: To debunk the myth that heat loss comes mostly from the head. The whole passage is dedicated to debunking the head heat loss myth. The passage says that "each part of your body loses its proportional amount of heat in accordance with its surface area," which means an area such as the chest would lose more heat than the head because it's bigger. The other answer choices are incorrect.

21. C: The primary topic of this passage is the use of serious themes in children's literature. The passage does say that children's literature holds a special place in many people's hearts; it does not say that it is universally loved by everyone. The passage does not mention that it is only to be enjoyed by children. Happy endings are mentioned but only in passing to prove a larger point.

22. C: This passage is providing information about children's literature and the use of serious themes within the genre. There is also no language to suggest it is a compare and contrast type passage. It does not just seek to describe or persuade.

23. B: The word "gruesome" means horrible and grim. In the passage, this word is used to describe stories that contain violence and death. This would not apply to comfortable, numb, or peculiar.

24. D: The original stories were grimmer and darker than more recent adaptations. The passage does not discuss the stories being translated from other languages or formats. While the themes from the original stories were more mature, there is no suggestion that they were intended for adults.

25. B: This statement is an example of figurative language called personification, where a thing or an animal has human characteristics. In this example, the stories speak to the children. The passage discusses how children interpret stories, so the correct answer is Choice *B*, children derive meanings from stories that adults do not.

Math Achievement

1. A: $\frac{3}{100}$. Each digit to the left of the decimal point represents a higher multiple of 10 and each digit to the right of the decimal point represents a quotient of a higher multiple of 10 for the divisor. The first digit to the right of the decimal point is equal to the value ÷ 10. The second digit to the right of the decimal point is equal to the value ÷ (10 × 10), or the value ÷ 100.

2. D: 13,078. The power of 10 by which a digit is multiplied corresponds with the number of zeros following the digit when expressing its value in standard form. Therefore:

$$(1 \times 10^4) + (3 \times 10^3) + (7 \times 10^1) + (8 \times 10^0)$$

$$10{,}000 + 3{,}000 + 70 + 8 = 13{,}078$$

3. C: 847.90. The hundredths place value is located two digits to the right of the decimal point (the digit 9 in the original number). The digit to the right of the place value is examined to decide whether to round up or keep the digit. In this case, the digit 6 is 5 or greater so the hundredth place is rounded up. When rounding up, if the digit to be increased is a 9, the digit to its left is increased by one and the digit in the desired place value is made a zero. Therefore, the number is rounded to 847.90.

4. C: $\frac{11}{15}$. Fractions must have like denominators to be added. We are trying to add a fraction with a denominator of 3 to a fraction with a denominator of 5, so we have to convert both fractions to their respective equivalent fractions that have a common denominator. The common denominator is the least common multiple (LCM) of the two original denominators. In this case, the LCM is 15, so both fractions should be changed to equivalent fractions with a denominator of 15. To determine the numerator of the new fraction, the old numerator is multiplied by the same number by which the old denominator is multiplied to obtain the new denominator. For the fraction $\frac{2}{5}$, multiplying both the numerator and denominator by 3 produces $\frac{6}{15}$. When fractions have like denominators, they are added by adding the numerators and keeping the denominator the same:

$$\frac{5}{15} + \frac{6}{15} = \frac{11}{15}$$

5. B: 34. When performing calculations consisting of more than one operation, the order of operations should be followed: *P*arenthesis, *E*xponents, *M*ultiplication/*D*ivision, *A*ddition/*S*ubtraction. Parenthesis:

$$7^2 - 3 \times (4 + 2) + 15 \div 5$$

$$7^2 - 3 \times (6) + 15 \div 5$$

Exponents:

$$7^2 - 3 \times 6 + 15 \div 5$$

$$49 - 3 \times 6 + 15 \div 5$$

Multiplication/Division (from left to right):

$$49 - 3 \times 6 + 15 \div 5 = 49 - 18 + 3$$

Addition/Subtraction (from left to right): $49 - 18 + 3 = 34$.

6. B: 80%. To convert a fraction to a percent, the fraction is first converted to a decimal. To do so, the numerator is divided by the denominator: $4 \div 5 = 0.8$. To convert a decimal to a percent, the number is multiplied by 100:

$$0.8 \times 100 = 80\%$$

7. C: 80 min. To solve the problem, a proportion is written consisting of ratios comparing distance and time. One way to set up the proportion is:

$$\frac{3}{48} = \frac{5}{x} \left(\frac{distance}{time} = \frac{distance}{time} \right)$$

$$\frac{3}{48} = \frac{5}{x} \left(\frac{distance}{time} = \frac{distance}{time} \right)$$

where *x* represents the unknown value of time.

To solve a proportion, the ratios are cross-multiplied:

$$(3)(x) = (5)(48) \rightarrow 3x = 240$$

The equation is solved by isolating the variable, or dividing by 3 on both sides, to produce $x = 80$.

8. D: 1, 2, 3, 4, 6, 12. A given number divides evenly by each of its factors to produce an integer (no decimals). The number 5, 7, 8, 9, 10, 11 (and their opposites) do not divide evenly into 12. Therefore, these numbers are not factors.

9. A: 11. To determine the number of houses that can fit on the street, the length of the street is divided by the width of each house: $345 \div 30 = 11.5$. Although the mathematical calculation of 11.5 is correct, this answer is not reasonable. Half of a house cannot be built, so the company will need to either build 11 or 12 houses. Since the width of 12 houses (360 feet) will extend past the length of the street, only 11 houses can be built.

10. C: The number line is divided into increments of .1 or $\frac{1}{10}$. The dot is located between the lines indicating .4 and .5. Choice *A*, $\frac{1}{4}$, would be .25, which is lower than .4, so it is incorrect. Choice *B* is incorrect because $\frac{2}{5}$ is equivalent to .4, which is not between .4 and .5. The number represented cannot be Choice *D*, because the dot is between .4 and .5, not on .5. Choice C, .45, falls between .4 and .5, so this is the correct answer.

11. D: 180 miles. The rate, 60 miles per hour, and time, 3 hours, are given for the scenario. To determine the distance traveled, the given values for the rate (*r*) and time (*t*) are substituted into the distance formula and evaluated:

$$d = r \times t$$

$$d = \left(\frac{60mi}{h}\right) \times (3h)$$

$$d = 180mi$$

12. D: $x \leq -5$. When solving a linear equation or inequality:

Distribution is performed if necessary:

$$-3(x+4)$$

$$-3x - 12 \geq x + 8$$

This means that any like terms on the same side of the equation/inequality are combined.

The equation/inequality is manipulated to get the variable on one side. In this case, subtracting *x* from both sides produces:

$$-4x - 12 \geq 8$$

The variable is isolated using inverse operations to undo addition/subtraction. Adding 12 to both sides produces:

$$-4x \geq 20$$

The variable is isolated using inverse operations to undo multiplication/division. Remember if dividing by a negative number, the relationship of the inequality reverses, so the sign is flipped. In this case, dividing by -4 on both sides produces $x \leq -5$.

13. C: $y = 40x + 300$. In this scenario, the variables are the number of sales and Karen's weekly pay. The weekly pay depends on the number of sales. Therefore, weekly pay is the dependent variable (y), and the number of sales is the independent variable (*x*). Each pair of values from the table can be written as an ordered pair (*x*, *y*): (2, 380), (7, 580), (4, 460), (8, 620). The ordered pairs can be substituted into the equations to see which creates true statements (both sides equal) for each pair.

Even if one ordered pair produces equal values for a given equation, the other three ordered pairs must be checked. The only equation which is true for all four ordered pairs is $y = 40x + 300$.

$$380 = 40(2) + 300 \rightarrow 380 = 380$$

$$580 = 40(7) + 300 \rightarrow 580 = 580$$

$$460 = 40(4) + 300 \rightarrow 460 = 460$$

$$620 = 40(8) + 300 \rightarrow 620 = 620$$

14. D: $x \geq 1$. The closed dot on one indicates that the value is included in the set. The arrow pointing right indicates that numbers greater than one (numbers get larger to the right) are included in the set. Therefore, the set includes numbers greater than or equal to one, which can be written as $x \geq 1$.

15. B: First, the number of employees who did not choose a hamburger must be found. If 5 employees chose a salad and 12 chose a sandwich, then $5 + 12 = 17$ did not choose a hamburger. This number can be subtracted from the total to find out how many employees chose a hamburger:

$$25 - 17 = 8$$

16. B: The average amount Cindy spent on groceries in March can be calculated by adding all the purchases together and dividing by the total number of purchases.

$$\frac{\$42.36 + \$26.50 + \$31.71 + \$37.23}{4} = \$34.45$$

17. B: Cone. A polygon is a closed two-dimensional figure consisting of three or more sides. A decagon is a polygon with 10 sides. A rhombus is a polygon with 4 sides. A triangle is a polygon with three sides. A cone is a three-dimensional figure and is classified as a solid.

18. C: If Joan travels for 2.5 hours with a 15 minute, or .25 hour stop, her total travel time is 2.75 hours. If she leaves at 10:15 a.m., she should arrive 2.75 hours later which would be 1 p.m.

19. C: 216cm. Because area is a two-dimensional measurement, the dimensions are multiplied by a scale that is squared to determine the scale of the corresponding areas. The dimensions of the rectangle are multiplied by a scale of 3. Therefore, the area is multiplied by a scale of 3^2 (which is equal to 9):

$$24cm \times 9 = 216cm$$

20. A: (-3, 2). The coordinates of a point are written as an ordered pair (*x*, *y*). To determine the *x*-coordinate, a line is traced directly above or below the point until reaching the *x*-axis. This step notes the value on the *x*-axis. In this case, the *x*-coordinate is -3. To determine the *y*-coordinate, a line is traced directly to the right or left of the point until reaching the y-axis, which notes the value on the *y*-axis. In this case, the *y*-coordinate is 2. Therefore, the ordered pair is written (-3, 2).

21. D: Perimeter is found by calculating the sum of all sides of the polygon. $9 + 9 + 9 + 8 + 8 + s = 56$, where *s* is the missing side length. Therefore, 43 plus the missing side length is equal to 56. The missing side length is 13 cm.

22. D: $\frac{1}{3}$ of the shirts sold were patterned. Therefore, $1 - \frac{1}{3} = \frac{2}{3}$ of the shirts sold were solid. Anytime "of" a quantity appears in a word problem, multiplication should be used. Therefore:

$$192 \times \frac{2}{3} = \frac{192 \times 2}{3} = \frac{384}{3} = 128 \text{ solid shirts were sold}$$

The entire expression is $192 \times \left(1 - \frac{1}{3}\right)$.

23. A: Mean. An outlier is a data value that is either far above or far below the majority of values in a sample set. The mean is the average of all the values in the set. In a small sample set, a very high or very low number could drastically change the average of the data points. Outliers will have no more of an effect on the median (the middle value when arranged from lowest to highest) than any other value above or below the median. If the same outlier does not repeat, outliers will have no effect on the mode (value that repeats most often).

24. C: Line graph. The scenario involves data consisting of two variables, month and stock value. Box plots display data consisting of values for one variable. Therefore, a box plot is not an appropriate choice. Both line plots and circle graphs are used to display frequencies within categorical data. Neither can be used for the given scenario. Line graphs display two numerical variables on a coordinate grid and show trends among the variables.

25. B: $\frac{1}{12}$. The probability of picking the winner of the race is $\frac{1}{4}$, or $\left(\frac{number\ of\ favorable\ outcomes}{number\ of\ total\ outcomes}\right)$. Assuming the winner was picked on the first selection, three horses remain from which to choose the runner-up (these are dependent events). Therefore, the probability of picking the runner-up is $\frac{1}{3}$. To determine the probability of multiple events, the probability of each event is multiplied:

$$\frac{1}{4} \times \frac{1}{3} = \frac{1}{12}$$

26. D: 3 must be multiplied times $27\frac{3}{4}$. In order to easily do this, the mixed number should be converted into an improper fraction.

$$27\frac{3}{4} = 27 \times 4 + \frac{3}{4} = \frac{111}{4}$$

Therefore, Denver had approximately $3 \times \frac{111}{4} = \frac{333}{4}$ inches of snow. The improper fraction can be converted back into a mixed number through division.

$$\frac{333}{4} = 83\frac{1}{4} \text{ inches}$$

27. C: According to order of operations, multiplication and division must be completed first from left to right. Then, addition and subtraction is completed from left to right. Therefore:

$$9 \times 9 \div 9 + 9 - 9 \div 9$$

$$81 \div 9 + 9 - 9 \div 9$$

$$9 + 9 - 9 \div 9$$

$$9 + 9 - 1$$

$$18 - 1 = 17$$

28. B: Using the conversion rate, multiply the projected weight loss of 25 lb by 0.45 $\frac{kg}{lb}$ to get the amount in kilograms (11.25 kg).

29. C First, subtract $1437 from $2334.50 to find Johnny's monthly savings; this equals $897.50. Then, multiply this amount by 3 to find out how much he will have (in three months) before he pays for his vacation: this equals $2692.50. Finally, subtract the cost of the vacation ($1750) from this amount to find how much Johnny will have left: $942.50.

30. B: To solve this correctly, keep in mind the order of operations with the mnemonic PEMDAS (Please Excuse My Dear Aunt Sally). This stands for Parentheses, Exponents, Multiplication, Division, Addition, Subtraction. Taking it step by step, solve inside the parentheses first:

$$4 \times 7 + (4)^2 \div 2$$

Then, apply the exponent:

$$4 \times 7 + 16 \div 2$$

Multiplication and division are both performed next:

$$28 + 8 = 36$$

Addition and subtraction are done last. The solution is 36.

Dear ISEE Lower Level Test Taker,

We would like to start by thanking you for purchasing this practice test book for your ISEE Lower Level exam. We hope that we exceeded your expectations.

We strive to make our practice questions as similar as possible to what you will encounter on test day. With that being said, if you found something that you feel was not up to your standards, please send us an email and let us know.

We would also like to let you know about another book in our catalog that may interest you.

SSAT Lower Level

This can be found on Amazon: amazon.com/dp/1628454237

We have study guides in a wide variety of fields. If the one you are looking for isn't listed above, then try searching for it on Amazon or send us an email.

Thanks Again and Happy Testing!
Product Development Team
info@studyguideteam.com

Interested in buying more than 10 copies of our product? Contact us about bulk discounts:

bulkorders@studyguideteam.com

FREE Test Taking Tips DVD Offer

To help us better serve you, we have developed a Test Taking Tips DVD that we would like to give you for FREE. **This DVD covers world-class test taking tips that you can use to be even more successful when you are taking your test.**

All that we ask is that you email us your feedback about your study guide. Please let us know what you thought about it – whether that is good, bad or indifferent.

To get your **FREE Test Taking Tips DVD**, email freedvd@studyguideteam.com with "FREE DVD" in the subject line and the following information in the body of the email:

a. The title of your study guide.

b. Your product rating on a scale of 1-5, with 5 being the highest rating.

c. Your feedback about the study guide. What did you think of it?

d. Your full name and shipping address to send your free DVD.

If you have any questions or concerns, please don't hesitate to contact us at freedvd@studyguideteam.com.

Thanks again!

CPSIA information can be obtained
at www.ICGtesting.com
Printed in the USA
BVHW010300101121
621267BV00013B/421